Marketing Your Farm

Marketing *Your* Farm

A practical guide to attracting loyal customers and increasing sales

MYRNA GREENFIELD

SAVVY PRESS
Salem, NY USA

ISBN 978-1-939113-60-3 (print) ISBN 978-1-939113-62-7 (ebook)
Library of Congress Control Number: 2022944443

Printed in the United States of America
First printing October 2022
Savvy Press, Salem, NY USA
savvypress.com

Designed by Sue Niewiarowski

The information in this book is true and complete to the best of our knowledge. While every precaution has been taken in the preparation of this book, the author and publisher expressly disclaim any responsibility for any errors, omissions, or adverse effects arising from the use or application of the information contained within. No liability is assumed for damages that may esult from the use of information contained within. This book does not offer legal advice. Always consult with an attorney for local laws, codes, and regulations.

Acknowledgements

This book is dedicated to:

My spouse, Michelle Johnson, who lovingly humors my passion for local food and farms, cooks delicious meals, and lets me drag her to a Pick-Your-Own farm on my birthday every year.

My book designer, copy editor, and long-time friend and colleague, Sue Niewiarowski.

My book coach, Leigh Belanger, who kept me going when I needed the encouragement.

People who have read and commented on various drafts of the book, including Adam Gibbons (Small Biz Helper), Alexis Breiteneicher Schneeflock (Valley Community Development), Apolo J. Cátala (OASIS on Ballou Farm), Dee Dice (Constant Growth Marketing and Consulting), Ellen Larson (Savvy Press), Mari Omland (Green Mountain Girls Farm), Kate Raisz (42 Degrees North Media), Rebecca Sadlowski (Rooted Flowers Farm), Jim Schultz (Red Shirt Farm), Julia Shanks (Julia Shanks Food Consulting), Samantha Staebner (Whittier Farms), Maida Tilchen (Savvy Press), and Rebecca Toms (UConn Extension).

The farmers and colleagues who have shared their wisdom, expertise, and experience.

Friends and family who have cheered me along the way.

Farmers and lovers of local food everywhere.

CONTENTS

DIG IN

"When solving problems, dig at the roots instead of just hacking at the leaves." ANTHONY J. D'ANGELO

TOPICS

- Welcome, farmer!
- Farm sizes used in this book
- Apply what fits for your farm
- KNOW, PLAN, DO, MEASURE
- My love affair with farms

Welcome, Farmer!

Your farm is unique because it's a reflection of you: Who you are, what you offer, and how, where, and why you farm.

Most direct farm customers are hungry for something more than a good head of cauliflower, even if they don't realize it. They want to know where their food comes from, support farms, and be part of a community with shared values.

Selling directly adds value to the products, services, and experiences you offer customers. How can you capitalize on that added value to help your business thrive? That's what this book's all about.

Over the years, I've worked with all kinds of farms, from beginning farmers running a CSA on a quarter-acre plot to multigeneration family farms adapting to a changing market. Here are their most common marketing questions:

- How can I increase sales?
- How can I get more people to find my farm online?
- How can I find and keep loyal customers?
- How can I make my farm stand out?
- How do I know if my marketing is working?

I can't give you a blanket answer to these questions because farm marketing isn't one-size-fits-all. The best tactics for your farm depend on its size, goals, years of operation, target customers, market conditions, and available resources (time, money, and talent). This book combines general advice for all farms with specific recommendations based on farm size. Apply what you read to your situation.

The best tactics for your farm depend on its size, goals, years of operation, target customers, market conditions, and available resources (time, money, and talent).

Farm sizes used in this book

Microfarms

This book categorizes you as a microfarm if you farm five acres or less, sell between $1,000 and $49,999 in agricultural products per year, and/or have been in existence for up to five years.

The USDA defines a small farm as any place that produces and sells $1,000–$349,999 of agricultural products annually. In my experience, however, there's a big difference between a $30,000 farm and a $300,000 farm. So I created a separate segment for microfarms.

If you're a microfarm, it's probably just you—or your household—farming. One or both of you may have an off-farm job. You're experimenting to determine which products, services, and experiences are profitable. You may be selling through one or two channels, such as a farmers market or an honor-system (aka self-serve) farmstand.

It's all you can do to grow, harvest, produce and sell. You feel like you *should* do some marketing, but you have so little time! Throughout the book, I suggest the *microfarm musts* that are most likely to help you grow your farm.

Small farms

This book defines a small farm as a business that grosses between $50,000 and $349,999 annually.

If you're a small farm, you've probably been in existence at least five years. You're likely to sell through three or more channels, which may include farmers markets, CSA, farmstands, retail stores, and restaurants. You may employ up to five people. You or someone else in your household may rely on some off-farm income.

You think you've got farming down pretty well. You know you need to do more marketing to increase sales and be more profitable, but you have limited resources and aren't sure where to focus. Follow the microfarm musts that are relevant to your farm and adopt as many of the *small farm smarts* as you can.

Midsize farms

This book defines a midsize farm as having gross farm income of $350,000 to $999,999 a year, in keeping with USDA's definition.

If you're a midsize farm, you've probably been around at least ten years—or much longer, if you're a multigeneration family farm. You may have a few or even dozens of family members or employees working with you.

Well-established midsize farms tend to be diversified; you rely on a mix of wholesale and direct-to-consumer sales. You may have a farm store offering a variety of products beyond what you grow or produce. You may offer some agritourism experiences, such as events and PYO[1]. You may also produce value-added products or sell online.

Midsize farms face increasing competition and price pressure. You need to modernize and further diversify to ensure that your farm will be around for the next generation.

You're probably already doing a lot of marketing, so consider my midsize farm suggestions as *maybes*. Read the microfarm musts and small farm smarts to see which ones make sense for your farm.

Before you take on more marketing, make sure you're measuring whether your tactics are achieving your goals. You may be able to do *less* marketing if you focus on the right things for your farm.

Apply what fits for your farm

Although every farm is unique, farms of similar sizes face similar challenges and are likely to benefit from using similar tactics. Therefore, many of the recommendations are organized according to farm size.

Throughout the book, you'll see *Where to focus* sidebars suggesting helpful content or tips for each farm size (see sidebar for an example). If there isn't a *Where to focus* sidebar, that section is useful for all farms.

Not sure which size fits your farm? Read the advice for all three sizes and choose the tactics that make sense for you.

Where to focus

MICROFARM MUSTS
Follow the recommendations for microfarms.

SMALL FARM SMARTS
Read and act on the microfarm recommendations that apply to your farm. Adopt as many small farm suggestions as possible.

MIDSIZE FARM MAYBES
Consider the advice for microfarms and small farms. Study the tactics for midsize farms and use the ones that apply to your farm's goals and needs.

1. Farms that let customers pick their own fruit, vegetables, or flowers are variously known as Pick-Your-Own, PYO, You-Pick, or U-Pick farms. The term PYO is used throughout this book.

How to use the *How-to* exercises

Scan any *How-to* boxes as you read the entire section. After you finish reading about a topic, go back to the *How-to* box. Decide whether this exercise will be useful.

Set aside some uninterrupted time to complete the exercise. Most *How-to* exercises can be completed in 10–15 minutes.

If writing is difficult for you, record and transcribe your response.

If you're the sole owner/operator of your farm, do this exercise yourself. If you farm with others, you may want to include them in this process.

If you get stuck, come back to the exercise later. Review what you wrote with a fresh eye. Finish if you can. If it still doesn't feel right, ask someone else at your farm for help.

The book also includes some *How-to* exercises that provide step-by-step advice for generating ideas, choosing tactics, or creating marketing content for your farm. *How-to* exercises appear in boxes with light purple backgrounds, like the one at the top of this column.

KNOW, PLAN, DO, MEASURE

Marketing Your Farm is organized into four parts: **KNOW, PLAN, DO,** and **MEASURE.**

1. **KNOW** covers the five essential aspects of farm marketing:

- What is marketing
- The 6Ps of farm marketing
- Know your market
- Know your customers
- Know your peers/competitors

2. **PLAN** guides you through the process of creating a marketing plan:

- How to turn a bunch of tactics into a plan
- How to create a speed marketing plan
- How to create a comprehensive marketing plan
- How to implement your plan

3. DO offers how-to advice for how to achieve specific goals:

- Build a strong brand
- Sell the sizzle
- Grow customer loyalty
- Increase sales
- Find new direct customers
- Find and keep wholesale customers
- Grow your website
- Grow your social media
- Get found online

4. MEASURE shows you how to:

- Track
- Measure
- Evaluate if your tactics are working

That's me, showing off a potato I just dug at Wards Berry Farm in Sharon, MA. (Photo: Michelle Johnson)

Each part is self-contained, so you can jump right into a topic or refer back to the book when a new need arises. I do encourage you to read the entire book—the overview will help you make better decisions about where to invest your marketing time and money.

My love affair with farms

Over the past four decades, I've marketed for businesses and nonprofits of all types and sizes. In 2010, I hatched Good Egg Marketing because I wanted to use my marketing expertise to promote the farms and food businesses that I loved. Good Egg's services include naming, branding, market research, marketing plans, sales coaching, websites, email marketing, social media—the works.

Since then, I've worked with micro-, small-, and midsize farms and food producers selling meat, fish, cheese, honey, maple syrup, fruit, vegetables, wool, tea, value-added products, personal care products, and more. I've marketed practically everything that farms produce, except marijuana—but I expect that to come soon.

I've taught marketing workshops to over a thousand farmers just like you. Many farmers get excited while I'm talking, but don't adopt any new marketing tactics afterwards. I get it. There aren't enough hours in the day for farming—when are you supposed to have time

I wrote this book because I love farms and want you, the farmer, to succeed—both for your sake and my own.

for marketing? And how the heck are you supposed to figure out which tactics to use?

I wrote this book because I love farms and want you, the farmer, to succeed—both for your sake and my own:

- I want you to keep producing fresh, delicious, healthy food and other products
- I want you to preserve farmland so that birds, bees, and endangered species can continue to pollinate and procreate
- I want you to build healthy soil and protect the environment
- I want you to be profitable and keep money circulating in your community
- I want you to thrive because your farm is essential to the world I want to live in

My connection to farming started in college. I spent my junior year studying at the University of Sussex near Brighton, England. I fell in love with the open-air markets where farmers hawked their fresh lettuce, tomatoes, and cheeses. I couldn't believe how wonderful fresh local food tasted, compared to the canned and frozen vegetables I'd grown up eating.

I joined Worldwide Opportunities on Organic Farms (WWOOF) so I could spend time in the bucolic English countryside. WWOOF sent me to a farm that had more volunteers than tasks, so I—a suburban kid who'd never grown so much as a tomato—was assigned to scrape dried cow dung from the barn's cement floor.

The experience squashed any desire to become a farmer, but it did help me appreciate the back-breaking labor that goes into growing food. And it didn't stop me from wanting to visit farms whenever I can.

Enough about me. Let's dig in!

KNOW

"The beautiful thing about learning is that nobody can take it away from you." B.B. KING

TOPICS

- What is marketing, anyway?
- The 6 Ps of farm marketing
- Know your market
- Know your customers
- Know your peers/competitors

What you'll find in KNOW

Where to focus

MICROFARM MUSTS

Read *What is marketing, anyway?*, *The 6 Ps of farm marketing*, and *Know your customers.*

SMALL FARM SMARTS

Read all the above, plus *Know your peers/competitors.*

MIDSIZE FARM MAYBES

Read all five sections.

KNOW covers the five essential aspects of farm marketing:

What *is* marketing, anyway? This whole book is about marketing—so why bother reading this section? It helps you understand what marketing is, why it's essential to the long-term viability of your farm, and why you need to keep adjusting your tactics to keep up with what's going on in the world.

The *6 Ps* of farm marketing. This section helps you analyze every aspect of your farm through a marketing lens. We'll look at your offerings, pricing strategy, distribution, presentation, people, and promotion. You'll learn how to leverage the *6 Ps* of farm marketing to generate a profit while staying true to your values.

Know your market. Your farm doesn't exist in isolation—it's part of one or more larger markets offering similar goods, services, and experiences. This section will help you analyze your market, understand who buys local food, and learn how to keep current with trends.

Know your customers. Understanding your customers is essential to building loyalty and attracting new people to your farm. This section helps you identify what you already know about your customers, find out what you need to know, and learn how to get the information you need.

Know your peers/competitors. Whether you see other farmers as your peers or your competitors, understanding how your farm compares to similar farms will help you identify and promote what's special about your farm. This section shows you how to position your farm based on your offerings and values.

What *is* marketing, anyway?

Marketing is one of those words that everyone uses, but hardly anyone can define. When some people hear marketing, they think *sales,* while others see sales and marketing as totally separate disciplines.

I define marketing as a set of processes, tools, and tactics you can use to achieve your business goals, including sales. It includes any actions you take to identify and reach out to target customers, stimulate sales, build customer loyalty, increase awareness of your farm, share your story, or other goals.

For example, if you want to increase sales at a farmers market, you could do one or all of the following:

- Grow or produce more of your bestselling items from previous years
- Create eye-catching signage and displays
- Promote that you're a veteran-owned farm
- Apply to accept SNAP payments
- Send customers a preview of what you'll be selling at that week's market, including recipes, photos, and storage information

All of the above activities are marketing tactics.

People often confuse marketing strategies and tactics. A *marketing strategy* is the overall approach you'll take to achieve a specific goal. *Marketing tactics* are the actions you'll take to implement the strategy. Each tactic may require multiple steps to implement.

For example, if your goal is to increase sales, your strategy might be to increase the average amount that customers spend when they visit your farmstand. Your tactics might be to offer more prepared foods and gift items.

If you choose tactics designed to achieve a specific goal, your strategy will be implicit in the actions you take. To keep it simple, this book doesn't ask you to map out a strategy before choosing tactics; it shows you how to choose the right marketing tactics for your goals.

Marketing is essential to your farm's viability

The Unknown Unknowns

Former US Secretary of Defense Donald Rumsfeld famously said that there are *known knowns* (things that we know that we know), *known unknowns* (things that we know that we don't know), and the *unknown unknowns* (things that we don't know that we don't know).

Operating a thriving, resilient farm is challenging in the best of times. There are many *known unknowns,* like climate change, that require constant adaptation. Farming is also full of *unknown unknowns,* like COVID-19.

You must assess your situation and adapt as needed. Marketing tools help you understand what's going on in the market, identify opportunities, and collect data and feedback so you can make more informed business decisions.

Marketing is as fundamental to your farm's success as producing quality products and maintaining soil health. You can't farm for a living without mapping out what you'll grow or produce, what labor and inputs you'll need, and how much you'll produce. Similarly, you can't run a profitable farm without matching what you offer with what your customers desire. Apply the same care to your marketing as you do to planning your crops.

You're already doing more marketing than you think

When most people hear the word marketing, they think about promotional stuff, like having a cool logo or posting to social media. Actually, marketing encompasses everything you do to create and deliver the products, services, and intangibles that satisfy your customers—and keep them coming back for more.

Deciding what to sell, where to sell it, what it should cost, and how to package or present it are all aspects of marketing. These elements can have as much of an impact on your farm's profitability as any of the promotional stuff. By identifying the marketing-related tasks you're already doing, you'll be able to coordinate them with your emails, website, and any other new promotional tactics you adopt.

If you're already spending time and money on marketing, why not do it as strategically as possible?

Your marketing tactics must keep up with the times

Although the essence of marketing hasn't changed in millennia, new tactics, tools, and technology to facilitate the process are always evolving. Whether you're a seasoned marketer or new to this stuff, you must stay up-to-date.

For example, color printing used to be expensive but is now relatively affordable. Advertising used to cost too much for most farms to consider, but now there are lots of ways to advertise inexpensively. These days, there are so many free and low-cost ways to connect with your customers, build your brand, and increase sales that it's easy to feel overwhelmed by the possibilities.

There are so many free and low-cost ways to connect with your customers, build your brand, and increase sales that it's easy to feel overwhelmed by the possibilities.

Today's customers want to use today's technology. Whether they're Zoomers or Boomers, your customers email, text, and post on social media all day long. They expect to reach you the same ways that they reach any other business.

Sales channels are changing. The places where you sold your products before may not be a good match for your farm now. The COVID-19 pandemic pushed many consumers to order groceries online for onsite pickup or home delivery and many farms set up websites to meet the demand. When you change sales channels, you should review your marketing tactics to see if they still make sense.

You can do this!

The art of marketing involves understanding your customers.

- What do they love about your farm?
- What motivates them to keep coming back?
- What are their values and dreams? How does your farm help fulfill them?
- What ideas, words or images will resonate with them?

The science of marketing involves analysis.

- What characteristics do your customers share?
- What trends are coming down the road?
- Which marketing tactics seem like a good match for your goals?
- How can you measure if your tactics are working?

While most people are more comfortable with either art or science, you must learn to do some of each. Start with the simpler tactics and build your skills over time.

The 6 Ps of farm marketing

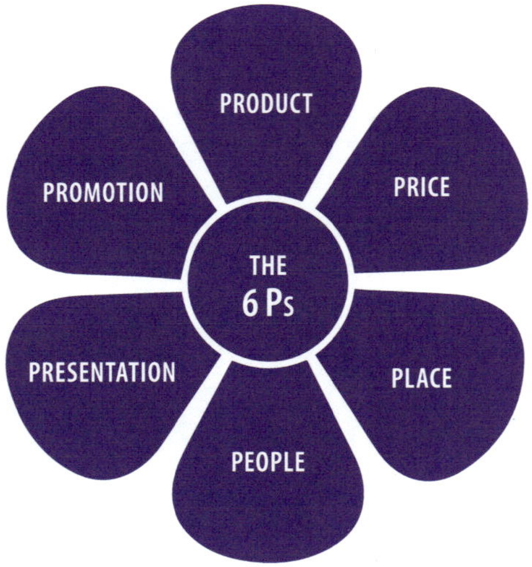

In 1960, marketing professor E. Jerome McCarthy published a textbook called *Basic Marketing: A Managerial Approach*. He identified the four major elements of marketing—Product, Price, Place, and Promotion—and dubbed them the "4 Ps."

These 4 Ps were such a big hit with marketers that most of us still rely on them today. When it comes to farm marketing, however, 4 Ps aren't enough. We need to add another two more Ps to the mix—People and Presentation.

Each of the 6 Ps is a building block of marketing that you can customize to identify and address your farm's needs and goals. Like other aspects of marketing, these elements are interdependent and ever-shifting. They can be developed and deployed in any order. Combining them effectively will help you get the best results from your marketing time and money.

The 6 Ps of farm marketing

1. **Products, services and/or experiences:** What do you offer?
2. **Price:** How does your prices compare to similar farms?
3. **Place (Distribution):** Where and how are your products, services, and experiences available?
4. **People:** How are you engaging with your employees; current and potential customers; and peers/competitors, suppliers, and partners?
5. **Presentation:** What impressions get conveyed when people see and interact with your farm, brand, and products?
6. **Promotion:** How are you getting the word out about your farm and offerings?

Use the 6 Ps to create a thriving, resilient farm

Farms of all sizes can benefit from conducting a 6 Ps brainstorm when addressing a current need, planning for upcoming seasons, or choosing a new opportunity. Spending a few minutes generating ideas will help you identify and choose the best options to achieve your goals.

How to use the 6 Ps framework

Identify one or more goals. Sample goals include reduce costs, diversify your revenue streams, or increase your CSA retention rate.

List events or market trends that might impact your ability to achieve each goal. For example, if many of your customers have temporarily relocated after a wildfire, you may need new distribution sites.

List changes you could make to achieve each goal. For example, if you're a PYO orchard that's been losing sales due to increased competition, how could you change your product, price, distribution, people, presentation, or promotion to make up for the lost revenue?

Brainstorm ideas for each P.

Highlight the most promising concepts.

Consider how your top concepts could impact each of the Ps. For example, if you're considering creating a premium line (Price) of Greek yogurt (Product), is it likely to expand your customer base or will the price scare away your existing customers (People)? What if you brought the higher-end yogurt to a new channel (Place), such as juice bars? What type of package design would you need to justify the higher price (Presentation)? Would you need to pay for advertisements (Promotion)?

Choose which idea(s) to explore further. Unless you're ready to make a major change at your farm, focus on small things that can have a big impact.

Gather the information you need to decide.

Go for it!

If possible, go through this process with at least one other person. When I run workshops teaching farmers how to use the 6 Ps, each participant brain- storms individually, then everyone shares and comments on one another's ideas. Feedback is like fertilizer—it yields richer results.

When weighing options, avoid jumping on the first idea that comes to your head. There may be an easier, cheaper, or more appropriate way to get the results you need. The best choice may be a combination or sequence of tactics.

For example, before deciding to increase revenue by selling at an additional farmers market (Place), consider other options. You could create a more eye-catching display (Presentation), improve your customer service (People), or use text messaging (Promotion) to attract customers to your existing markets.

Reviewing the 6 Ps will help you define the core elements of your farm. Once you've clarified the 6 Ps, it will be easier to develop a strong brand that reflects your farm's offerings, values, and personality.

Conducting virtual events

When Eden Specialty Ciders in Newport, VT had to suspend in-person tastings during the pandemic, they began offering virtual cider tastings via Zoom.

"Over the course of six months, we held 30 public tastings and 20 tastings for private groups," co-owner Eleanor Leger says. "It added 10% to our revenue and generated new memberships and add-on sales."

Although Eden has resumed in-person cider tastings, it continues to offer virtual public and private events because they're profitable.

Products

What products/services/experiences do you offer?

The mix of products, services, and/or experiences that you offer provides the most tangible benefits to your customers and helps differentiate your farm. Adding new offerings isn't always the best way to achieve your goals; adapting or subtracting an offering may yield better results.

Could you build on something you already do? If you're already growing diversified vegetables, adding an unusual new crop may not require much effort, whereas getting into a new distribution channel could be a much bigger enterprise. Similarly, once you provide one agritourism offering, it may be easy to add another farm tour.

Products. A *product* is usually something tangible that can be eaten, used, consumed, or downloaded, but it could also be a concept or even a nonfungible token (NFT). What about herbal and medicinal products, convenience items, non-edibles, flowers, compost, energy production, or marijuana?

Services. A *service* is something that you do or offer on behalf of your customer. It can include a product as part of the offer, but the service component must add value for the customer. Potential services include any form of customer assistance, delivery, planting and landscape design, floristry, catering, specialty CSAs, accommodations, or photography/videography.

Experiences. An *experience* is a distinct, memorable, usually pleasant event or activity that your customer attends or encounters. It may rouse physical, sensory, or intangible reactions. It may be a stand-alone experience or incorporate one or more of your products or services. Potential experiences include agritourism, food service, recreation, fairs, concerts, hands-on farm activities, contests, games, online offerings, or virtual reality.

Price

Balancing affordability with your farm's viability

The essential work that you and other farmers do—growing food, producing products, preserving agricultural land, creating jobs, and supporting local economies—is severely undervalued. US farmers receive 15 cents of each dollar American consumers spend on food, down from 37 cents in the 1980s. Yet many Americans don't have access to nutritious, affordable, and culturally appropriate food. Consolidation in the food supply chain is benefiting a handful of food conglomerates and their shareholders, not farmers and eaters.

We need a US food policy that addresses both farmers' and eaters' needs. Until we have such a policy, find a pricing strategy that enables you to meet your goals, pay yourself and staff a living wage, and put money aside for emergencies and retirement. If the market won't support the prices you need, try to collaborate with food hubs, cooperatives, and nonprofits that offer support.

Promoting SNAP benefits

Mona Suimay sells produce from her grandmother's farm, Song Farm, at Copley Square Farmers Market in Boston, MA. Their SNAP benefits banner is prominently displayed.
(Photo: Myrna Greenfield)

When it comes to pricing, context matters

Your farm exists in a local, state, regional, national, and global context. You're competing against every other business that can satisfy the same needs as your farm. You're also operating within a larger economy that affects peoples' ability and desire to purchase your products, services, or experiences.

Commodity prices are relatively *elastic*; when prices go up, demand tends to go down. If the price is too high, consumers will buy something else or go without. If your products are seen as interchangeable commodities—such as produce, meat, dairy, or Christmas trees—you may have to sell at market rate. When consumers will buy a product almost regardless of how much it costs, then the price is relatively *inelastic*, meaning that demand will stay about the same, even if the price goes up.

Red Fire Farm's certified organic tomatoes at Copley Square Farmers Market.
(Photo: Myrna Greenfield)

Certified organic heirloom tomatoes were selling for $6.50 a pound at Red Fire Farm at Copley Square Farmers Market in Boston, MA in early summer; Whole Foods was selling organic tomatoes on the vine for $3.49 that same week.

Capturing a higher price

Farmers typically charge similar prices for similar products at farmers markets. If your products are unique, available earlier in the season, or offer other special characteristics, you can get a higher price.

The same principles apply to services or experiences, such as farm stays, PYO, catering, petting zoos, festivals, etc. The more interchangeable and easily available, the more price sensitive it's likely to be. If you offer a service or experience that's seen as more unique or higher quality, you'll be able to claim a higher price.

You can usually command a premium just by selling direct because customers assume products that come straight from a farm are fresher and/or of higher quality. Look for ways to differentiate your products, services, or experiences so they'll be perceived as being worth more. For example, you can offer special sizes or flavors, include an experiential component, provide more convenience, be more environmentally responsible, or package or present your products in an eye-catching way.

Pricing strategies

Businesses typically use one or more of the following methods to set prices: *Cost plus, competitive,* and *value-based* pricing.

Cost-plus pricing involves calculating your fixed and variable costs and adding the amount you need to earn your desired profit. Some farms calculate costs for *each* product they grow or produce, while others group costs by category (such as vegetables, fruit, or dairy) or enterprise (such as PYO, farmstand, or bakery).

Competitive pricing is based on what farms or other businesses that produce similar products are charging. You can set pricing at mid-range (somewhere between the lowest and highest prices of your competitors), above-market (higher than your competitors), or below-market (cheaper than anyone else).

Value-based pricing is based on the *perceived* value that your products, services, or experiences offer for your target customers. You price your offerings based on what they're worth to your customers, not your costs of production or what your competitors are charging. If you can offer your customers a product, service, or experience that meet their needs better than anyone else, many will pay whatever it costs.

Place (Distribution)

Distribution matters. The right distribution channels can help you increase sales and become more resilient. For example, farms that were exclusively selling to restaurants or institutions pre-COVID-19 got caught in terrible straits when food service came to a standstill during the pandemic. Dairy farms that didn't already sell to stores or direct to consumers had nowhere to sell their milk. While most farms eventually found new distribution outlets, there was a lot of crying over spilt milk before the situation was resolved.

In many cases, you can diversify your operations by expanding an existing service rather than by finding a new distribution channel. During the pandemic, many farms began offering online ordering for shipping, delivery, or pickup at farmers markets or other outlets. Some farms teamed up with other farms and local food businesses to provide at-home delivery, while other farms signed up to sell through existing local food delivery services.

As the demand for local and/or organic food has increased, wholesale opportunities have also expanded. Many farms do best by selling to small independent foods stores, coops, and regional chains that like to buy direct. As you scale up, you may want larger opportunities. Most retail outlets, such as supermarkets, and chains, as well as mass market stores and wholesale clubs (Walmart, Target, Costco, Sam's Club), operate on extremely low margins, but you may be able to make this work for your farm.

When it comes to distribution options, look for synergy. If you already have a CSA, adding a delivery service might aid customer retention. If you're wholesaling to institutions, adding new institutions might be easier than trying ecommerce. Look for ways to leverage existing relationships to find new distribution outlets.

Creating more value

Green Mountain Girls Farm in Northfield, Vermont is a diversified farm that offers deluxe farm stays. "Our prices are high because we feel that we're offering something relatively rare: An authentic farm stay," says co-owner Mari Omland. "There are farm-based accommodations that don't give visitors a chance to directly experience life on the farm. Families that stay here have a custom experience that gets them as close to the crops, the animals, and daily life on the farm as they want to be." She adds, "People who want a luxury hotel are willing to pay for it. And so are people who want a taste of authentic farm life."

People

You, your family, and employees

The people who work for your farm can help you attract customers just as much—or more—as the products, services, and experiences you offer. Everything they say and do helps you to live your brand and express your values. When your staff feel valued, they're likely to reflect that back to your customers.

Encourage positive interactions between your staff and customers. Simple things—like giving employees farm-branded clothing or accessories to wear—will help your brand stand out and let customers know whom to ask when they need help.

If you're a family farm, be sure to involve the next generation in decision-making. For example, when Barstow's Dairy Store and Bakery in Hadley, MA realized they needed to expand beyond dairy in order to sustain the farm, the stimulus to start a bakery came from one of the younger family members with a passion for baking.

Your current and potential customers

Having loyal customers is key to your resilience. You can diversify by seeking to attract new types of customers. You can target foodies, families, people who follow specific diets, or any group that has dietary needs or interests that you can serve.

Your partners, peers, and suppliers

Your business relationships can help you diversify and differentiate your farm. When you partner with other farms who grow things you don't produce, you can support other farmers and provide more one-stop shopping for your customers. When you sell products and services to Black and Brown businesses owners, you can support their businesses. If you import products that aren't grown in the United States, you can support small producers by buying directly from other farmers or sourcing through Fair Trade organizations.

A growing number of consumers—especially millennials and Gen Zers—choose to shop with socially responsible businesses. If you intentionally use business relationships to support your values, share what you do and why with your customers.

ECO-packaging by Red Tomato (Photo: Red Tomato)

Presentation

Presentation includes signage, displays, merchandising, packaging, overall look-and-feel, and branding. Presentation primarily encompasses visuals, but could include any way that you represent your farm in person, in print, digitally, or through an experience.

For example, *unboxing*, or the experience of opening a package, has become something that many people consider part of a product's value. Many businesses include a little thank you note or a free gift in the box they use to fulfill ecommerce orders.

Red Tomato—a Northeast food distribution nonprofit based in Providence, RI—markets and distributes *righteous produce*™. They designed an innovative, breathable plastic package that draws attention to its ECO apple® line, while noting the specific farm where the apples are sourced.

Promotion

Promotion encompasses any tactics you use to attract attention, build awareness of, generate interest in, or spur sales for your farm, products, services, and experiences. Promotional tactics include signage, advertising, public relations, social media, sponsorship, endorsements, events, couponing, and stunts.

Promotion is essential for attracting and retaining customers. In addition to drawing attention to your farm, it can educate customers about your offerings, suggests new ways to use products, express your values, and differentiate your farm.

Where to focus

MICROFARM MUSTS

Understanding the market will help you make better decisions when you want to expand your farm. Read *Why you need to know your market, The US food market,* and *The US local food market.*

SMALL FARM SMARTS

Read all the above. If you sell to retailers, institutions, or distributors, also read *Market, market area, category, market size, and market share.*

MIDSIZE FARM MAYBES

Read all the above. Research any markets or categories that are important to your business.

KNOW YOUR MARKET

Why you need to know your market

Your farm, products, services, and experiences are competing for sales in a local, regional, national, and even global market. Markets are dynamic. New trends, products, and businesses pop up all the time, so you need to keep your eyes open.

Whether you peddle produce at a farmers market, sell salsa to stores, or rent your barn for weddings, learning about the larger context for your offerings will help you understand your customers and identify opportunities and risks for your farm.

The US food market

The US food industry is split between retail food outlets and food service. US consumers spend slightly than half their food budget on food they eat away from home. This includes a wide range of food service outlets, including national chain and local independent restaurants, food trucks, cafeterias, sports venues, food trucks, and vending machines.

The other roughly half of their food dollar is spent at retail food outlets, including supermarkets, convenience stores, specialty food shops, and discount stores. Supermarkets, warehouse clubs, and supercenters capture the vast majority (85%) of those purchases.

The COVID-19 pandemic created some shifts in the market. When it forced food service operations to shut down in March 2020, there was an uptick in grocery sales and cooking at home. By March 2021, spending on food outside the home returned to pre-pandemic levels.

Consumers report, however, that they're continuing to cook at home more often than they did before the pandemic. Retail food sales bear this trend out. For example, the value of retail food sales during the week ending May 1, 2022, was higher than during the same week in 2019 (although inflation accounted for a portion of those sales).[2]

2. USDA Economic Research Service. (2022, June 30). *Sales at retail food stores were higher in May 2022 than May 2019, but growth varied by food category.* https://www.ers.usda.gov/data-products/chart-gallery/gallery/chart-detail/?chartId=104184

The pandemic also prompted many US shoppers to buy groceries online. Only 6% of US grocery sales took place through ecommerce in 2019; by 2021, online shopping accounted for 9.6% of US grocery sales.

While some online shoppers have gone back to brick-and-mortar stores, ecommerce is projected to capture over 20% of the market by 2026. If you offer a delivery service—or are considering it—be aware that the majority of consumers prefer curbside or in-store pick-up to home delivery. These preferences vary by location, so be sure to ask or survey your customers about their pick-up preferences.

Traditional and emerging drivers of food shopping

Traditionally, consumers weigh three factors when shopping for food: **Quality, price,** and **convenience.**

Quality. Quality is a deal-breaker for shoppers; they won't buy a product or service if they don't trust the quality. The average consumer can't assess quality the way that you, chefs, nutritionists, or produce buyers can, so they look for characteristics that symbolize quality to them. They judge products based on perceived freshness, flavor, appearance, aroma, growing methods, variety, and other aspects. **Make sure your offerings emphasize quality by keeping your displays immaculate and using language and images that project freshness and flavor.**

Price. Price is a factor for almost all shoppers. People on limited budgets examine the price of every item. Some shop at stores like Dollar Stores or Walmart that they perceive to have more affordable prices. SNAP (food stamp) users must go to outlets that accepts EBT payments. Even people who can afford to spend as much money as they want may refuse to purchase a product or service if it seems too expensive for "what it should be." At the same time, shoppers often assume that if something costs more, it must be higher quality. **Consider offering products at a variety of price points so you can serve a broader range of consumers.**

Convenience. All consumers—rich and poor, young and old—are driven by convenience. Saving time is probably the most important convenience factor for shoppers. Other factors include location, hours, store layout, check-out time, ease of getting through aisles, parking, transportation, and delivery. **Look for ways to add convenience to help differentiate your offerings.**

Emerging concerns. In addition to quality, price, and convenience, consumers are increasingly weighing additional factors—such as health and wellness, safety, social impact, experience, and transparency—when making purchasing decisions. If your offerings are strong in any of these areas, emphasize these characteristics in your signage, packaging, and other communications.

The US local food market

Direct sales of local food—including both fresh and value-added foods—accounts for a tiny portion of the US food market. In 2020, 147,000+ US farms produced and sold $9 billion of local edible food commodities (a 3% increase from 2015).[3] To put this in perspective, $9 billion is less than 0.5% of the $1.71 trillion worth of food that the US foodservice and food retailing industries sold that same year.[4]

One-third (33%) was sold directly to consumers at outlets such as farmers markets, onsite farm stores, roadside stands, CSAs, online sales, PYO, and mobile markets. Just two of those channels—farm stores and farmers markets—accounted for 59% of direct-to-consumer sales. Sales in this category decreased slightly, dropping from $3.0 billion in 2015 to $2.9 billion in 2020.

One-fifth (21%) was sold directly to retailers, including supermarkets, supercenters, restaurants, caterers, independent grocery stores, and food cooperatives. This figure dropped from $2.3 billion in 2015 to $1.9 billion in sales in 2020.

---

3. USDA. (2022, April). *Direct Farm Sales of Food, 2017 Census of Agriculture Highlights, Direct Farm Sales of Food, Results from the 2020 Local Food Marketing Practices Survey.* https://www.nass.usda.gov/Publications/Highlights/2022/local-foods.pdf

4. USDA Economic Research Service. (2022, July 22). *Food Service Industry, Market Segments.* https://www.ers.usda.gov/topics/food-markets-prices/food-service-industry/market-segments/

Nearly half (46%) of direct farm sales went to intermediate markets, including institutions (schools, colleges, universities, and hospitals) and intermediary businesses (wholesalers, distributors, and processors). Intermediate market sales *increased* from $3.4 billion in 2015 to $4.1 billion in 2020.

What can we conclude from this data? Remember, COVID-19 disrupted the market in 2020, so we can't make too many generalizations. However, sales to institutions and intermediary businesses made up the largest portion of direct farm sales in both 2015 and 2020, so that trend is likely to continue. Clearly, selling to institutions and intermediary businesses is a growing market opportunity for farms to explore.

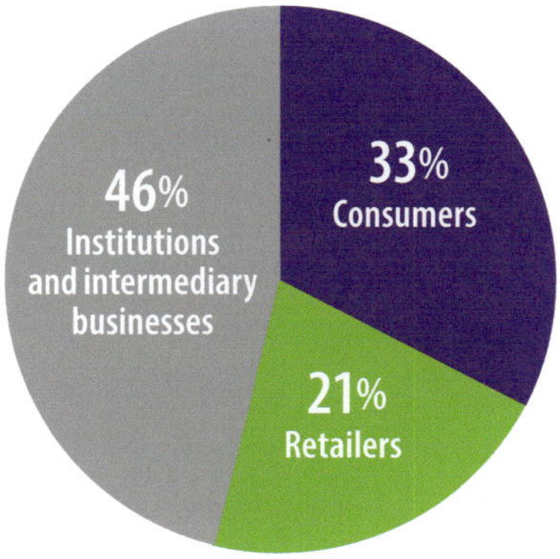

Direct farm sales of food by type of buyer

(2020 sales in billions)

Who buys local food?

Consumers can buy local food directly from you or through retailers, institutions, or intermediaries. What factors influence whether they seek out local food and, if so, if they buy direct or through another channel?

What the research says

All types of people buy local food. Local food consumers transcend demographics; people of all sexes, age groups, races and ethnicities, income levels, education levels, and parts of the country buy local. Researchers have tried to determine if any demographic characteristics make people more likely to buy local food, but the results are inconclusive. People do have different needs and behaviors based on their age, household composition, income levels, and racial and ethnic backgrounds. **Researching the demographics of the locations you serve can help you target customer needs more effectively.**[5]

People don't always act on their beliefs. The most common reasons people cite for wanting to buy local food include superior freshness/nutrition, support for farmers and the local economy, and

5. Visit https://censusreporter.org and enter a zip code, city, or other location to get demographic information.

it's better for the environment. Yet studies show that people don't always follow through on these beliefs when deciding where to shop or what to buy. **When wooing customers, share your values, but also promote your quality, convenience, and affordability, as appropriate.**

Only a *few behaviors* matter. You might assume that people who follow a special diet or belong to a fitness club would be more likely to buy local food, but these factors don't have a significant impact on their purchasing habits. The behaviors that *do* increase the likelihood of seeking local food include enjoyment of cooking, gardening, shopping at a health food store, and buying organic food.

Treat your *committeds* like gold. Ask for feedback about your offerings and reward them for their loyalty.

The three types of local consumers

In my experience, the most effective way to target local foods customers is by their level of commitment to buying direct from farms. I categorize consumers into three groups: *Committeds, supportives,* and *indifferents.*

Committed to local. People who are committed to local (aka locavores) go out of their way to buy directly from farms whenever they can, even if it takes extra effort or costs more. When committeds do shop at a supermarket (or any indirect channel), they'll seek out local products, even if it means tracking down an employee or doing a quick online search to get the information they need. Despite their overall commitment to buying from farms, if your offerings are of lesser quality, unaffordable, or inconvenient, even committeds will go elsewhere at least some of the time. **Treat your committeds like gold. Ask for feedback about your offerings and reward them for their loyalty.**

Supportive of local. People who are supportive of local *want* to support local farms and the local economy, but are less likely to go out of their way to buy direct, especially if it takes more time or costs more. Supportives do the majority of their grocery shopping at a supermarket, natural foods store, or discount outlet, where they can get all their groceries at once, shop at almost any hour, park or get there easily, and buy at an acceptable price point. Products must be clearly identifiable as local, as supportives won't track down the information themselves. **Make sure your signage clearly identifies**

your farm's offerings; try to partner with retailers and intermediary businesses that prominently display and promote local products.

Indifferent to local. People who are indifferent to buying from local farms won't seek you out. There are many reasons why people don't buy local food. They may be struggling to make it through the day. They may not have the time, energy, or transportation to shop at more than one location. The products may be more expensive. They may prefer tropical or specialty foods that aren't locally grown or produced. Or they may have other values or priorities that are more important to them. **Don't try to persuade indifferents to shop with you; if their circumstances change, they'll come to you.**

These categories are fluid; customers may change their buying behavior over time. Unless you're sure that someone is committed, it's best to assume that they're supportive of local but need encouragement and incentives to become a loyal customer.

Market, market area, category, market size, and market share

What's your market?

The word *market* is an all-purpose term that can refer to:

The demand for or interest in a specific type of product, service, experience, or concept (local apples, a scooped ice cream shop, farm-to-table dinners, or regenerative agriculture).

The current or potential buyers/users of a specific offering who share similar interests or needs (parents of young children, college students, retirees, RVers, natural foods customers, luxury shoppers, or members of a specific immigrant community).

A geographic area (town, county, state, region, or country) where the offering is purchased or consumed.

A physical or online location (farmers market, supermarket, restaurant, hospital, ecommerce store) where an offering can be bought, acquired, or used.

In general, think of a market as the context in which your offering is desired, promoted, purchased, and/or used.

How to research markets

Start your internet research with broad search terms, so you can get a wide range of results. For example, if you're a certified organic grower, search for a term like "organic industry trends." If you're a cheesemaker, search for "specialty cheese market."

When researching more specialized topics, such as demand, pricing, or competitors, get as specific as possible. For example, instead of "heirloom wheat," search for "increasing demand for heirloom wheat in Midwest." If you can't find market data for your specific location, revert to broader national search terms.

Limit your search to results from the last 12 months so you get the most current information; extend the time period if necessary.

What's your market area?

Your *market area* refers to the geographic territory where you sell your products. If you sell within a 25-mile radius of your farm, your market area will be different than if you sell to an entire state, nationally, or internationally.

Your location and the size of your market area can impact your potential sales volume, pricing, competition, target consumers, and sales channels. If you farm, live, and sell in a single market area, you may already know it well, but if you sell to a broader area, you may benefit from researching it.

If you sell produce to local institutions *and* sell your value-added products nationally, you're operating in two different market areas. You'll need a separate marketing strategy for each of them.

What's your category?

A market *category* is a grouping of similar products, services, or offerings. If you offer more than one type of product or service, use more than one production method, or sell through more than one sales channel, your farm and products may fall into multiple categories.

Subcategories are small groupings of similar products or services within a broader category. At the broadest level, your category is a farm; your subcategories may be dairy, livestock, produce, or Christmas trees. Your sub-subcategory may be grass-fed beef and so on. See the sample category breakdown to the right.

Farm
Livestock
Grass-fed beef
Organic grass-fed beef

When categories matter

If you sell directly to consumers and don't do much marketing, categories aren't relevant for your farm. If you sell to retailers, institutions, and intermediaries, knowing your category can help you:

- Gather information about your market, customers, pricing, product usage, and peers/competitors
- Position your offerings to appeal to consumers in your category
- Persuade store buyers for your category to carry your products

When consumers are making decisions about what products, services, or offerings to buy, they're comparing your offerings to similar items in your category. **The smaller your category, the easier it is for your product to stand out.** If consumers view the soft cheese you make from your dairy as just "cheese," you're likely to be one of many cheeses under consideration. If they view your category as "local cheese," you'll probably be one of a select few. It will be easier for you to convey the added value you offer and be able to sell at a higher price point.

You don't need to place everything you offer into separate categories. For example, if you grow spinach and Swiss chard, you probably sell these products to similar customers in similar ways, so it's not worth defining yourself as a spinach farmer *and* a Swiss chard farmer. If you wholesale your apples *and* bake and sell apple pies, you're selling to two different types of customers, each with different needs, so it makes sense to group them into separate categories.

Selling your products, services, or offerings by category

Most categories are standard. There are some variations from store to store, especially when new types of products are brought to market.

Stores maintain departments (such as produce, meat, grocery) and group products by categories within each department section. Buyers determine the volume, vendors, brands, and flavors to stock by category. Many buyers conduct periodic "category reviews" to determine which products and vendors to keep and which to drop.

When Beyond Meat launched their plant-based burgers, they were seeking to broaden the appeal of their products beyond the usual consumers of vegetarian products. In order to get on the radar of "flexitarians"—eaters who enjoy both meat and vegetarian options—they thought it was crucial to get their products into the meat aisle. The Kroger and Safeway supermarket chains agreed to sell the product in the meat category, where it became a hit. So categories matter.

How to calculate market size

If you can't get the exact data you need for your city or state, you can do a little math to get an estimate for your target area.

Say you want to know the total number of cage-free eggs consumed in your town each year. An internet search will tell you that the per capita US egg consumption for 2020 was an estimated 286.5 eggs/year[6] and that 22.5% of eggs produced are cage-free.[7]

If you know the population in your target area, you'd multiply it by 286.5 x 22.5% to get an approximate number.

What's your market size?

Market size can either refer to the total sales volume or total number of customers who purchase products, services, or experiences in a particular category.

Your market size will depend on how you define your category and your market area. Are you measuring the market size for all yogurt, organic yogurt, or organic maple syrup yogurt? Are you measuring sales for the entire US, your region, your state, or your town?

Knowing your approximate market size is helpful when you're trying to decide whether to get into a new market. You usually don't need a precise figure to plan and budget. Knowing the exact size of your market *can* be important if you're writing a business plan or going after funding.

For-profit market research firms create detailed reports that can cost thousands of dollars. If you can't afford to purchase a report, search for articles, press releases, or blog posts that share the

6. Statistica. (2022, January.) *Per capita consumption of eggs in the United States from 2000 to 2021.* https://www.statista.com/statistics/183678/per-capita-consumption-of-eggs-in-the-us-since-2000/

7. Statistica. (2021, March). *Distribution of laying hens population in the U.S. in 2020, by management type.* https://www.statista.com/statistics/995596/share-of-laying-hens-in-the-us-by-management/

highlights. In addition, you may be able to get access to reports through a local, university, or business library.

What's your market share?

Market share refers to the portion of the total market your farm is capturing. For example, the Massachusetts cranberry market sold $59,652,000 of product in 2017, according to the 2017 USDA Census of Agriculture. If your farm sold $59,652 in cranberries in 2017, you captured 0.1% of the Massachusetts cranberry market.

For most farms, producing a great product, service, or experience and selling it to appreciative customers is a better path to success than worrying about your share of the market. If you're selling value-added products to retailers, however, knowing your market share can help strengthen your pitch to store buyers.

How to assess market share

Visit stores that carry your product category and count how many other farms or businesses are selling the same or similar products at each store. Who has the most shelf space? Who gets the best placement and promotion? If you're one of three farmers or producers and you all seem to be getting about the same amount of space on the shelves or produce racks, you can guestimate that you've each got a 33% market share in that location.

Big companies have access to data from all of their sales outlets, so they can keep tabs on everyone's market share. If you sell to large stores like Whole Foods or through distributors, you should be able to get similar data.

Trend-watching

Trend-watching is basically just a way to pay attention to what's going on in the world. If you're trying to increase sales or expand, it's a great way to get ideas for new offerings, flavors, and tactics.

Tools for staying on top of trends

Your sales data. If you track sales of your offerings, you should be able to do a year-over-year analysis of sales for each item. If you sell through wholesale, the stores or distributors may make sales data

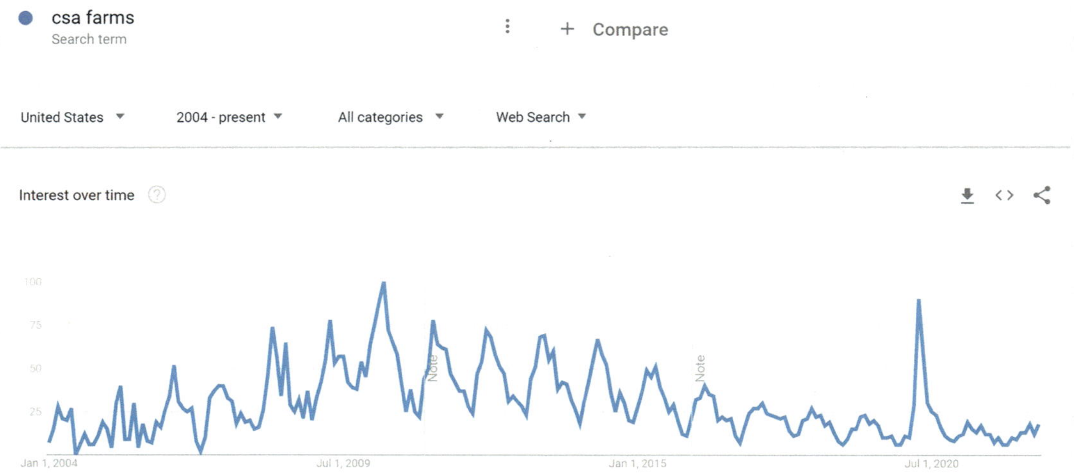

How to use Google Trends

Google Trends is my favorite way to get quick insights on an industry or product. Go to https://trends.google.com/trends/ and type in a word or phrase. Google will show you how many searches have been conducted.

The search posted above shows you how many times users in the United States searched for the phrase "CSA farm" between 2004 and 2022. Note the spike that occurred in April 2020, just after the COVID-19 pandemic hit. You can search by the entire US or your region, state, or town.

available to you. Are sales of any particular product or flavor increasing or decreasing? What are your customers asking for or commenting on?

Trend reports. Market research firms, media outlets, businesses, and social media channels publish annual trend reports summarizing everything from the most popular food delivery items to the color of the year.

Observation. Visit the places you sell to currently or hope to sell (stores, restaurants, farmers markets, or hospitals). Keep your eyes open when traveling to another part of the country. What varieties, flavors, and specialties are available? Do you see any unique items that aren't available in your market?

KNOW YOUR CUSTOMERS

It's all about your customers

Your customers are vital to your success. Studies show it costs five times more to attract a new customer than to nurture an existing one. Having a loyal customer base will provide you with a predictable income and help you use resources more effectively.

Figuring out who your customers are, what they need and want, and the best ways to communicate with them is essential to finding and keeping satisfied customers. If they love what you do, they're likely to encourage their friends and family to support you.

The more you know about your customers, the easier it will be to keep them happy and to find more customers like them. If you can satisfy their needs, you may be able to sell more to them by promoting your existing offerings; adding additional products, services, or experiences; or deepening their loyalty.

You may already know a lot about your current customers from observing and interacting with them. This section shares simple techniques for understanding how to gather information about your customers. You'll also learn how to create a customer chart and/or profiles for each type of customer you serve.

If you want to scale up your farm or start a new enterprise, you may also need to conduct surveys, interviews or focus groups to get more detailed information.

Where to focus

MICROFARM MUSTS

Read *The three types of customer information you need.*

SMALL FARM SMARTS

Read *The three types of customer information you need and Create customer profiles.* Use this information to help you target your offerings and messages to each type of customer you serve.

MIDSIZE FARM MAYBES

Read all the sections, create a customer chart, do some research (if needed), and segment your customers by needs and behavior, so you can reach them with the right offerings and messages.

The three types of customer information you need

There are three types of information that can help you better understand and satisfy your customers.

Demographics. Demographic information is factual data about your customers, such as where they live or work, their age, sex, and household size.

Needs, values, and interests. This area covers what's going inside your customers' heads and hearts. What are their inner needs and motivations? What do they believe? What do they value? What are their hobbies and interests?

Satisfying your customers' needs and wants

Let's look at a local foods customer: Me. Why do I shop at farmers markets? I'm a foodie who will go out of my way for the freshest, tastiest, and most unusual foods I can find. I also shop at farmers markets because I want to support farmers and local vendors, I like being around other people who are excited about food and farms, and it makes me happy.

When I visit Copley Square Farmers Market in Boston, MA, I always stop by the Red Fire Farm booth, because they offer an amazing bounty of products. Their artful displays remind me of the markets I frequented as a student in England.

Shopping at farmers markets satisfies a physical need to buy groceries *and* fulfills my emotional needs.

Red Fire Farm display at Copley Square Farmers Market in Boston, MA. (Photo: Myrna Greenfield)

Behaviors. This refers to tangible, measurable actions that your customers and visitors take, such as attending an event at your farm, shopping at your farmstand, joining a CSA, sharing a Facebook post, watching a video, or navigating your website.

Demographics

Knowing where your current customers live or their typical life stage can help you plan where, what, and how to sell to them.

Location

Location can refer to where your customers live, work, or shop. Your customers may draw from a wider geographic area than you realize. For example, if you offer coffee and donuts or grab-n-go meals at your farm store, you may attract people passing by on their way to or from school or work. Your PYO customers may come for a yearly outing from several hours away.

With the advent of food delivery and shipping, local farms can choose to sell to customers in a broader range of locations than in the past.

How to obtain location information

If you have a POS, take credit card payments, or use loyalty cards, run a report to determine how many customers you have in each town or zip code.

Get data from your web site or social media. Most social media channels and website platforms will show you how the number of visitors by location.

Ask. If you don't have access to location data through any of the above sources, ask people for their zip code when you make a sale.

Poll. If you have an email list or use social media, conduct a one-question poll to ask your subscribers or followers for their zip codes.[8]

Once you know how many customers you have in each town or zip code, choose which locations to target. Set up direct mailings, search engine marketing, and advertisements to reach current or potential customers in your target location.

Other useful demographics

Age

Knowing the age range of each type of customer can be useful. Although we all don't "act our age," each generation tends to share similar world views, priorities, and interests. For example, millennials (those born between 1981 and 1996) are more concerned about the social and environmental responsibility of the companies they do business with than other generations. This makes them likely to be interested in buying from local farms and businesses.

Use of social media also tends to vary by age group. Instagram is currently attracting users under 35 at a faster pace than Facebook, so if you want to reach younger people, you're better off focusing on Instagram than Facebook.[9]

---

8. Since only a portion of your followers are likely to respond, the responses may not be representative of your entire customer base, but you'll have some information to go by.

9. Social media preferences are always in flux; to get the latest information on which channels each generation uses, do an internet search for a phrase like "social media channels used by age."

Life stage

Our needs, interests, and buying behavior are more likely to be shaped by the stage of life we're currently experiencing than by our chronological age.

Here are the four life stages that are most relevant for farms:

- Young Singles
- Older Singles
- Young Marrieds (with children at home)
- Empty Nesters (children no longer at home)

Unless you conduct a survey, you won't know the precise age or life stage of your customers, but you can generalize based on appearance and behavior. If you have a cluster of customers at the same life stage, you can offer products and services likely to meet their needs. For example, Young Marrieds are likely to seek out activities like petting zoos or PYO, while Older Singles might be more attracted to wine and cheese tastings.

Needs, values, and interests

Assessing needs through observation

Steve Murray, owner of Heart Beets Farm in Berkley, MA, runs a vegetable CSA. Over time, he noticed that many first-time CSA members seemed overwhelmed by the unfamiliarity and sheer quantity of produce in their shares. He started steering new CSA members to sign-up for his biweekly "petite shares" instead of his larger weekly shares. The result? More first-timers are renewing their memberships for another year.

Needs and wants. When a person buys a product, service, or experience, they're seeking to fulfill a specific need or desire. A need is something that someone "must-have," whereas a want or desire is something that would be "nice-to-have." Needs and wants may be functional (something that fulfills a specific, tangible purpose), emotional (an intangible, internal desire that they may not even be aware of), or both.

For example, someone may buy your sausages at the farmers market because they need something for breakfast the next day (functional), but they may also seek you out because you're a regenerative farm that sells grass-fed meat (emotional desire to support their values).

Needs may be simple or complex. Some needs can be fulfilled by a simple sequence of actions (I'm hungry, I go to the refrigerator, I take out an apple, and I eat it), while others require many steps to resolve.

The amount of effort that we put into satisfying a need usually depends on how complex, important, or costly it is. Most of us put a lot more effort into buying a house than into choosing a snack!

There may be a gap between what we say we want and our true desires. People are often driven by multiple and/or competing motivations. For example, I may prefer shopping at farmers markets, but I'll go to the supermarket with a friend because I want to spend time with them.

There's no way that you can know, or act on, all of your customers' needs and wants. Harvard researchers have identified some 300 different emotional motivators![10] Meeting even just a few of your customers' wants can help you connect with them on a deeper level.

Values. Local foods shoppers are often motivated by the desire to fulfill their values, such as supporting the local economy and protecting the environment. Shopping or acting based on your values may be a conscious or unconscious motivation. The ideal customers for most farms tend to be values-driven. Highlight any values you share with your customers when you create messages and offers.

Interests. Interests, hobbies, and activities can motivate customers to shop with or visit your farm. For example, people who are interested in cooking or gardening are more likely to buy from local farms. If your farm is scenic, offers nature and recreational opportunities, specializes in producing a particular type of food or drink, holds concerts, or offers any type of activity likely to appeal to people with those interests, you can create offers, messages, and advertisements to attract them.

Behaviors

Behaviors are tangible actions that can be observed, collected, or recorded, such as how often a customer visits your farm store, how much they spend per visit, or how often they buy cheese.

Your ability to record and analyze customer behavior will depend on which automated or manual systems you use to run your business. For example, if you use a POS system or online ordering, you'll automatically be collecting data about your best-selling items, daily sales volume, and more.

10. Magids, S., Zorfas, A., & Leemon, D. (2015, November). The New Science of Customer Emotions, *Harvard Business Review*. https://hbr.org/2015/11/the-new-science-of-customer-emotions

How to review existing data sources

Sales data. Review your QuickBooks account, spreadsheets, or whatever systems you use to track sales. Look for trends, such as your best-selling categories, items, and services.

Customer information. Review any information that your POS or mobile sales app gathers information—such as how long someone has been a customer, how frequently they visit, and their zip code—that you can use to identify and target your most loyal customers

Digital analytics. If you have an email newsletter, texting service, loyalty program, or social media, review which content and offers get the most response.

Website analytics. If you offer ecommerce or have access to your website analytics, review which offerings, pages, and content attract the most sales or interest.

If a customer makes online purchases, has joined your loyalty program, signed up to receive your emails, or has otherwise consented for you to collect and use their data, you may be able to individually target individuals for specific offers or messages based on their behavior.

Create a customer chart

Once you've collected information about your customers' demographics, needs/values/interests, and behavior, put it together.

The sample customer chart to the right is a way to map groups of customers with similar characteristics and interests, so you can create offerings, messages, or ads targeted to each type of customer. It will also help you plan how to reach each group.

I recommend that you create your chart in a spreadsheet like Google Sheets or Excel, so you can easily add rows as needed. But you can draw it on a piece of paper or write it in a text document—whatever's easiest for you.

Sample Customer Chart

PRODUCT	WHERE/WHEN THEY BUY	WHO	WHY	HOW TO REACH THEM
Vegetable CSA	Pickup once a week in nearby city	Young marrieds, mostly women with disposable income who enjoy cooking	Want their families to eat healthy and know where their food is from; support farmers and local economy; impress others with their locavore credentials	At the pick-up site, email, social media, listserv for their city
Vegetable CSA	Pick up on your farm once a week	Young marrieds, mostly women with kids, mix of affluent and limited income families	Want to save money and preserve food to eat local year-round; support farmers and local economy; be part of a community of like-minded people	At your farm, email, social media, local events
PYO fruit	Seasonal, on farm	Young singles with disposable income	Have an experience, be in nature, enjoy the view, impress others with their locavore credentials, support farmers and local economy	Email, direct mail, social media, targeted social media or print ads
PYO fruit	Seasonal, on farm	Young marrieds with various incomes	Want healthy foods for their family; have an experience; be in nature and enjoy the view; pick food to preserve; support the local economy and farms	Email, direct mail, social media, targeted social media or print ads, Farm to School activities, partnerships with PTAs
Grass-fed lamb customers	Year round, at farmstand	Empty nesters, people from cultures that enjoy lamb; empty nesters who ate lamb as children	Important part of their food culture; nostalgia; comfort	At your farm, email, local events, partnerships with churches or organizations that serve people from people from that culture

Where to focus

MICROFARM MUSTS

If you sell to one type of customer, you can target them directly, so you probably don't need customer profiles.

SMALL FARM SMARTS

If you sell to more than one type of direct or wholesale customer, create customer profiles *or* a customer chart to help you target each type of customer more effectively.

MIDSIZE FARM MAYBES

If you're trying to increase sales for an existing or new offering, having profiles of your target customers will help you choose the right tactics, offers, and messages. If someone else—a staffer or consultant—does some or all of your marketing, customer profiles will help ensure that you're working with a shared understanding of your customers.

Create customer profiles

Most of us instinctively think of a specific person when we hear the word "customer." You may not even know his or her name, but you know that they buy a pie every Saturday morning or always share Facebook videos of your farm animals.

A customer profile (often called a *customer persona* in marketing jargon) describes a typical customer with specific needs and wants that you fulfill. Customer profiles are typically composites of many customers with similar problems or desires, rather than a description of a single actual customer.

It's easier to create effective offers, messages, and communications when you can picture a specific person and how they're likely to react. When you're drafting an email or creating an ad, you can write as though you're speaking directly to that person.

Let's look at the sample customer profile below. What can we observe about PYO Parents? They're young parents, so need convenience and efficiency. They're enjoying their time on your farm. They're a Black family.

How can you attract PYO Parents based on this bio? Offer family-friendly activities, such as hayrides, a petting zoo, and music. Set up online PYO scheduling, so there's no wait on busy weekends. Post or advertise your farm on listservs that the Black community follows.

Sample Customer Profile

PYO Parents

Bio: Busy parents with young child

Pain point: Want their child to learn about food and nature, but it's hard to schedule family outings

Product/service/experience: Fall outing for PYO pumpkins

Why buy from you: Family-friendly and culturally diverse setting

(Photo: Depositphotos)

How to create a customer profile

Profiles are a tool to help you picture, communicate with, and create products, services, or offers that will connect with your target customers. Only include the information that will help you do this. Try to incorporate the exact words or phrases that customers say about your farm into your profile.

Typical elements for customer profiles

Name. Create a descriptive name for your customer to make it easier to remember their profile. Pretty soon, you'll start associating the profiles with your actual customers. You'll spot a thirty-something woman with several young children in tow and think, "Here comes Mary the Millennial Mama!"

Examples:

- PYO Parents
- Greg the Gourmet
- Chef Cheryl
- Food Service Director Dave
- Farmers Market Fred

Photo (or avatar). Attaching a face with the profile name will help you get inside that person's head. Don't use a photo of an actual customer; it could get awkward if they ever happen to see it. Find a photo or avatar of someone who resembles the customer who you have in mind in terms of age range, gender, race or ethnicity (if relevant), clothing, etc.

Bio. List a few relevant points, such as age, sex, location, profession, and life stage. Sample bio: Farmers Market Fred is single, in his mid-40s, manages a pharmacy in your small rural town, and loves to cook dinner for friends in his home.

Need/want. What's the main need or want they are trying to address? (If you don't know, guess.) Sample need/want: Farmers Market Fred is an outgoing man who dreams of being a chef instead of managing the town pharmacy.

What/where/how they buy from your farm. Sample: Farmers Market Fred buys grass-fed meat and organic vegetables from you at the weekly farmers market on the town square (and indoors in winter).

Why they buy from you. What's the main reason they choose your farm to obtain this product or service? (If you don't know, guess.) Sample: Going to the market is the highlight of Farmers Market Fred's week. He loves the quality of the food you produce and your knowledge about how to prepare each cut. He enjoys telling you about what he cooked last week.

DIY customer research

After creating a customer chart or profile, you may need to fill gaps in your customer knowledge through a customer poll or survey, one-on-one interviews, or focus groups.

Customer polls and surveys

Polls. Customer polls are a quick, easy way to gather feedback about a single topic. You can create polls using your email program, Instagram Story, Facebook page, or other digital tools. In addition, you can easily embed a poll onto your website. Polls often get a good response because people like taking them.

Customer polls are a quick, easy way to gather feedback about a single topic.

Stick with simple yes or no questions, such as:

Are you potentially interested in getting delivery services from our farm?

☐ Yes

☐ No

Surveys. Surveys can be conducted in print, online, by phone, or in person, but most are conducted online because it's the easiest way to get responses. There are many free and inexpensive survey tools and some email service providers—such as Constant Contact and Mailchimp—also let you conduct polls and surveys. Use surveys to get feedback from CSA customers, to ask customers where they spend time online, or to test responses to an image, such as a new logo or salsa label.

Keep your surveys short (10 questions or less); people are less likely to complete longer surveys.

Ask one topic per question. For example, don't ask, "How long have you been a CSA member and what do you like about membership?" Make this two questions.

Offer an incentive to people who complete the survey.

Be cautious when interpreting responses. The people who answer surveys are likely to be your most loyal customers or—sometimes—the angriest ones. Unless your survey size is large enough, you can't assume that the responses are representative of your entire customer base.[11]

Even if just a few customers share a problem or need, you can use that information to improve your marketing messages, products, or services. For example, if you discover that your farmstand hours are inconvenient, you can stay open later one night per week. Or you might learn that your customers don't know what Integrated Pest Management (IPM) is, so you add an explanation on your website.

One-on-one interviews

One-on-one interviews are conversations with one person. One-on-ones give you a chance to ask in-depth questions and follow-up for additional detail. They allow you to listen for tone-of-voice or watch for body movements to gauge feelings and reactions. Sometimes a ten-minute talk can surface an issue that you'd never have thought of on your own. Prepare your questions in advance and record the conversation so you can review it later.

Focus groups

Focus groups are a great way to get a small group of people to share ideas or provide feedback. They typically consist of 6–8 people who gather in-person or via a video conference call for 1–2 hours. Small group discussions can help surface problems or solutions that may not come up when you're talking to one customer at a time.

Here are a few tips:

- Use open-ended questions
- Avoid sensitive issues like politics or religion
- Call on people. This encourages everyone to participate and helps you keep a know-it-all from dominating the group.

---

11. To figure out how many people you must survey to be confident that your responses are representative of an entire population, do an online search for *sample size calculator*.

Where to focus

MICROFARM MUSTS

Read *Competition matters, Competition can be good for your farm,* and *Who are your competitors?* Study the 2–3 closest or most similar farms and evaluate how your farm is different or better.

SMALL FARM SMARTS

Read the entire section. Identify the 2–3 closest or most similar farms for your top-selling offerings and evaluate how your farm is different or better.

MIDSIZE FARM MAYBES

Read the entire section. Identify the 2–3 closest or most similar farms for your top-selling offerings, create a chart of competitors, and write a positioning statement that shows how your farm is different or better.

KNOW YOUR PEERS/COMPETITORS

Competition matters

Most farmers get uncomfortable when I talk about competition, because you're usually friends with the farmer down the road. But whether you call other farms your friends, peers, or competitors, customers are making a choice between you and another farm.

The most effective way to compete is convey who you are, your offerings, and values, so your target customers can distinguish your farm from your peers/competitors.

Competition can be good for your farm

The conventional business world often portrays sales as winner-take-all, as if there's only room for one successful business in your market.

There are some situations where this is true. For example, Whole Foods allocates a finite amount of shelf space for bottled salsas. The buyer might have to get rid of someone else's product to make room for yours. Or if you're running a PYO operation, most customers will visit you or a neighboring orchard, not both.

Competitors can help build the market for your offerings, so there's more business for everyone. Having similar businesses nearby can drive traffic to your farm. Ever notice how often two coffee shops are located across the street from each other? You'd think that if there was already a coffee shop in a spot, it would be the worst location to open up. In fact, it's often the best spot; your competitor has already established the location as a good place to get a cup of coffee. You can capture the commuters who are traveling in the opposite direction or the people who are tired of waiting in line at the other shop.

When a bigger farm creates a market for a new category or product, it paves the way for smaller farms to compete. Being first to market has advantages, but it can be expensive to advertise or teach customers about a new product. For example, when farmers first started selling hydroponics, they had to educate their market. Now that more people are familiar with this growing method, you can skip the education and attract hydroponics customers by offering better tasting or more sustainably-produced products.

Who are your competitors?

Your competitors include any business that your customer might choose instead of your farm. Any business on the planet could theoretically be a competitor if they can meet the need or want that a consumer is trying to fulfill.

Having too many choices can be paralyzing, so consumers instinctively narrow their options as quickly as possible. As the famous "jam study" showed, consumers are more likely to buy jam when offered only 6 choices rather than 24 or 30 options.[12]

When someone has an immediate physical or emotional need, they'll probably choose the closest, quickest, or easiest way to satisfy it. If they have a more complex or important need, they'll probably consider more than one business or product. Serious buyers often create a list of selection criteria, such as quality, price, convenience, values, reputation, etc. They limit their options to a more manageable group, then make their purchase.

In marketing jargon, this narrowed-down list of options is called the customer's *consideration set*. I create a consideration set each time I choose a PYO farm for my birthday outing.

Rather than viewing everyone in the world as a direct competitor, focus on the businesses that your current or prospective customers are most likely to consider. If you sell directly to nearby consumers, your competition will be different than if you sell nationally (see top chart on next page). If you offer more than one type of product, service, or experience, or sell through a variety of locations or channels, your competitors for each offering will vary.

I'm following my bliss at a PYO flower farm. (Photo: Michelle Johnson)

Choosing a PYO farm

Every year, I visit a PYO farm to celebrate my birthday. How do I choose from among the 237 PYO farms in Massachusetts? I search online for farms that are less than two hours away and use organic, IPM, or regenerative growing methods.

Choosing a farm is low stakes for customers like me. I want the day to be special, but hey, there's always next year.

If your farm depends on PYO income, you must attract customers by expressing how you're different (or better). To do that, you must know your peers/competitors.

12. Iyengar, S. S., & Lepper, M. R. (2000). When choice is demotivating: Can one desire too much of a good thing? *Journal of Personality and Social Psychology, 79*(6), 995–1006. https://doi.org/10.1037/0022-3514.79.6.995

Sample Chart of Competitors for a Grass-fed Meat Farm

YOUR FARM	YOUR CUSTOMERS	YOUR COMPETITORS
A grass-fed meat farm who sells directly to consumers in a 50-mile radius	People who want to buy grass-fed meat directly from local farms	Other grass-fed meat farms in a 50-mile radius of your farm
A grass-fed meat farm that ships anywhere in the US	People who want to buy grass-fed meat, but it doesn't have to be local	Any outlet in the US that sells or ships grass-fed meat, including ecommerce stores, supermarkets, and Costco

Sample Competitors Table

SALES CHANNELS, OFFERINGS & SPECIALTIES	YOUR FARM	FARM #1	FARM #2
Certified organic produce	X		
Sell at farmers markets	X		X
CSA	X		X
Wholesale to institutions	X	X	
Additional specialty #1	Grass-fed meat sold at farmers markets and CSA	Grass-fed meat shipped via ecommerce	Grass-fed meat sold at farmers markets and CSA
Quality and pricing	Premium	Premium	Medium
Customer service	Excellent	Good	Okay
Target customers (if known)	Relationship-oriented foodies	Convenience- oriented foodies	Families with young children

The *Sample Competitors Table* above shows how your farm compares to two farms with similar offerings. Based on the information in the table, your farm is most similar to Farm #2. With your excellent customer service and top quality, you'll appeal to high-touch customers who enjoy lots of direct contact with their farmer.

You and Farm #2 both sell at farmers markets and have CSAs, but your products are certified organic, are premium quality, and are priced accordingly. Although you sell through some of the same channels, you occupy a different market position than Farm #2.

How to analyze your competitors

Focus on the competitors that your target customers are most likely to consider alongside your farm. If you view every entity that can theoretically meet your target customers' needs or desires as a competitor, you'll feel overwhelmed!

Your goal is to gather information so you can compare how you're similar and different from your competitors. This will help you develop messages highlighting your best features to your target customers.

Steps to follow

Identify other farms that provide similar offerings in your chosen market area.

- Ask your direct customers where else they shop or visit—and why
- Identify which farms rank at the top of search engine results for your offerings
- Check the websites, catalogs, or shelves of stores and institutions who buy your products to identify their other customers
- Ask the buyer at your wholesale customers to share perceptions of your competitors
- Visit locations where your products are sold to see which farms have a presence there

Create a spreadsheet or draw a table on a piece of paper.

- List your top 3 revenue producing activities in the left-hand column
- Add a few blank rows to list additional offerings from your competitors
- Enter the names of 2–10 competitor farms in the top row

Gather information about each farm from their websites and social media. How are they similar and different from you?

Place an x in the relevant columns for each farm.

Review the table. Star the 2–3 farms most similar to you and compare how you're similar and different.

The good way to compete

Once you've identified your competitors, focus on your strengths, not the competition. You don't have to criticize or malign other farms to get noticed. How do your products, services, and experiences satisfy your customers' needs and desires? How are your offerings different or better than your competitors?

When creating offers, messages, and advertisements for your farm, emphasize how you meet your customers' needs. For example, if you know that convenience is important to your target customers and you're competing with a supermarket, you could post a message like: "Pop by our farmstand on your way home to pick up a delicious heat-n'-eat dinner made with veggies from our farm." You're offering proof points that your products are convenient while emphasizing how you're different or better than the supermarket.

Create a positioning statement

Once you've figured out who your target customers are and what makes you special, you're ready to create a *positioning statement*. This helps you define how your farm meets the needs of your target customers. It emphasizes your strengths and points of differences and uses proof points to make your case.

A positioning statement is just for internal use. In most cases, you wouldn't want to publish it anywhere because it may sound boastful or jargony.

Positioning statements typically use a framework like this:

For people who are [insert the top products, values, and benefits you provide], [insert your farm name] is the [insert adjective such as best, only, leading, one of the best, etc.] choice, because we offer [insert 1–3 proof points].

Sample positioning statement:

Old Macdonald's Farm is a great CSA for families with young children, because we offer PYO opportunities for CSA members, children's entertainment during pickup hours, and a monthly potluck. In addition, we sell our organic produce to our local elementary school.

PLAN

"I skate to where the puck is going to be, not where it has been." WAYNE GRETSKY

TOPICS

- A marketing plan is your GPS
- Create a speed marketing plan
- Create a small farm marketing plan
- Create a comprehensive marketing plan
- Create a marketing implementation plan

What you'll find in PLAN

It's hard to make time for planning, so I've made this process as streamlined as possible. This part of the book gives you three options: A speed marketing plan, a small farm marketing plan, and a comprehensive marketing plan. It closes with tips on how to implement a marketing plan.

The speed marketing plan focuses on one goal at a time. The template asks you to set a goal, choose one type of customer to target, pick one or more tactics that can achieve your goal, and set the measurements you'll use to determine if you reached your goal. Any size farm can benefit from using this tool.

The small farm marketing plan uses the same format as the speed marketing plan, but allows for more detail and complexity. If you have multiple goals or want to target more than one type of customer, use the small farm marketing plan.

The comprehensive marketing plan is best suited for midsize farms or small farms that are rapidly expanding. It assesses your market, analyzes your current marketing tactics, recommends strategies and tactics to achieve your goals, and maps out tasks, timeline, budget and responsibilities. This is a big undertaking, so most farms only create a comprehensive plan when applying for funding.

Regardless of whether you actually create a plan, check out the *Create a marketing implementation plan* section to get ideas for streamlining your marketing tasks.

A marketing plan is your GPS: Don't farm without it

If you want to get better results or reduce the time and money you spend on marketing, you must be strategic about the tactics you use. A marketing plan sets goals and maps out the steps you'll take to achieve them. You do your best to choose effective tactics, but you won't know if they'll succeed until you try them. If they don't work, you'll try something else.

Your marketing plan is a GPS, not a printed map. If you get lost or hit a detour, a GPS will recalculate your route. Similarly, if you need to try a new tactic, your marketing plan will help you adjust course without having to start from scratch.

Create a speed marketing plan

I developed this speed marketing approach for farmers who don't have the time, expertise, or need to develop a full marketing plan. A speed marketing plan focuses on one aspect of your farm at a time. It enables you to try out a tactic and evaluate if it's working.

Here are the four components of a speed marketing plan:

1. **Set your goal:** What specific and measurable goal will you achieve and by what date?
2. **Choose your target customers:** What type of people are you trying to reach? To do what?
3. **Choose your tactic(s):** Which tactics will spur your target customers to take your desired action?
4. **Measure your results:** How will you know if it worked?

Where to focus

MICROFARM MUSTS

Use the speed marketing template to plan tactics for one aspect of your farm. It should take less than an hour.

SMALL FARM SMARTS

Use the small farm marketing template if you want to coordinate tactics for multiple goals. You should be able to draft your plan in 2–3 hours. If you've only got an hour, use the speed marketing template.

MIDSIZE FARM MAYBES

Use the speed or small farm marketing template when you need a quick plan for one part of your farm. If you're starting or scaling up a major enterprise, invest in creating a comprehensive plan. Read the *Create a marketing implementation plan* section for tips on how to coordinate your marketing tactics.

Sample Speed Marketing Plan

GOAL	TARGET CUSTOMER(S)	TACTIC(S)	MEASUREMENT
Attract 25 more customers to our annual seedling sale	People who shop in our town center	Distribute "buy 3 seedlings, get 1 free" coupons at local businesses	Number of customers and total sales this year comparied to last year Number of coupons redeemed Costs to print coupons and grow seedlings Net revenue this year compared to last year

STEP 1: SET GOAL

A speed marketing goal should be specific, measurable, and time-bound (achievable within a set time period).

STEP 2: CHOOSE TARGET CUSTOMERS

Your target customers are the people you must persuade to do something that will enable you to achieve your goal. In order to target them effectively, they must share at least one common characteristic (needs, motivations, location, demographics) and you must have a common way to reach them. When creating your first speed marketing plan, choose only one group of target customers per goal.

STEP 3: CHOOSE TACTICS

A tactic is something that you do, say, or offer to inspire your target market to do something that helps you achieve your goal. While some goals may require a series of tactics to implement, I recommend that you test one tactic at a time when creating your first plan.

STEP 4: MEASURE RESULTS

Record your current sales level or another relevant benchmark so you can measure progress from a starting point. If you don't have a system or tool in place that enables you to measure results, you'll need to set something up.

A speed marketing plan is a full circle that starts and ends with your goal. If you don't achieve your goal, you can change tactics and measure your results again. You can adapt your plan as you test tactics or conditions change.

Create a small farm marketing plan

This section follows a similar format as the speed marketing plan, but helps you achieve more complex goals. We'll use a dairy that produces specialty cheese as an example throughout this section.

Step 1: Set your goals

Keep asking yourself questions until you create a specific, measurable, and time-bound goal.[13]

How to Create a Specific Goal

INITIAL GOAL	MORE SPECIFIC GOAL	FINAL GOAL
Increase cheese sales	Increase annual cheese sales at farmers markets by 10%	Achieve a 10% annual increase in cheese sales by selling $500 of cheese at each of three farmers markets (A, B, and C) for 22 weeks

GET SPECIFIC

Ask yourself over what time period? Through which sales channels? By how much? Sales goals should ideally be based on sales data from a previous period. If you don't have that information, make an educated guess.

ADD INFORMATION

Ask yourself which farmers markets will I choose? How many weeks does each market run? How much cheese can I sell each market day?

Step 2: Choose your target customers

Target customers are a group of current or potential customers who share at least one thing in common. For example, they may all be passionate about cheese! In addition to sharing the same need or desire, you must be able to reach your target customers using the same communications channel. That could be in-person (at a weekly farmers market), by email, or through flyers at a local business.

13. While most farms use quantitative goals (something you can measure with a number), you can also set qualitative goals (something you can measure by gathering information). For example, if your goal is to "increase customer awareness of the full range of products, services, and experiences that we offer," you can use observations, surveys, or customer comments to assess if awareness has increased.

Sample Tactics

Free/low-cost	Higher cost
Email	Social media ads
Social media	Radio and TV ads
Partnership	Sponsorships
Calendar listings	Contests
Discounts	Events
Sampling/demos	Local chef demos

How to Choose Your Target Customers

INITIAL TARGET	FINAL TARGET
Cheese lovers	Cheese lovers who shop at farmers markets A, B, and C each week

GET SPECIFIC– Ask yourself what need do they share? What communication channel will I use to reach them?

Step 3: Develop your tactics

In Step 2, you chose a group of target customers with a similar need whom you can reach in the same way. In Step 3, you'll create a plan to persuade those customers to take your desired action.

Your first task is to come up with one or more tactics that will motivate them to act. A tactic can be an offer, a message, a display, a presentation, or any type of interaction—in person or digital—that captures your target customers' attention and persuades them to take a specific action,

In marketing jargon, the specific thing you're asking them to do is referred to as a *Call to Action* (CTA). A CTA is something specific and measurable that you want someone to do, such as buy something, click a link, share a post, or watch a video.

Sometimes you only need one tactic to get the results you need. For example, if you post a "Fresh Corn next exit" sign on the highway, customers are likely to show up!

You may need to employ a series of tactics in a particular sequence to achieve your goal. For example, if you want to attract more loyal customers, you must attract their attention, get them to make a purchase, motivate them to buy from you again, remind them to buy from you one more time, and so on, until buying from your farm becomes a habit.

Sample Small Farm Marketing Plan

GOAL	TARGET CUSTOMER(S)	TACTICS/RELATED TASKS	MEASUREMENT(S)
Achieve a 10% annual increase in cheese sales by selling $500 of cheese at each of 3 farmers markets (A, B, and C) for 22 weeks	Cheese lovers who shop at farmers markets A, B, and C each week	Create 4 sample Cheese Plate of the week platters Develop program logistics and costs Create a punch card or loyalty club for frequent buyers. Create a landing page with one-click subscription Create a unique promotional code for each communication Create a postcard and table signage Promote at the market and through social media Send weekly email Train staff to sell and promote the program Run a raffle at each market to collect emails for potential subscribers	Number of cheese plates sold per market/week Number of punch cards taken and number of punch cards redeemed Number of times promotional codes were used for each communication Engagement on social media posts Evaluation (written) of program and tactics: keep, modify, or end it

For our cheese example, our CTA will be "Buy our cheese plate of the week." The cheese plate will be a ready-to-eat spread featuring our cheeses and complementary items from other farms.

We'll need to map out details, such as creating sample cheese plates to test the concept, taking photos, calculating material and labor costs, designing a promotional postcard and table signage, creating a frequent buyer punch card or loyalty program, promoting via email and social media, and a raffle. (If this seems like too much, you can set more modest sales goals and see how your cheese plates sell without the additional promotion.)

Step 4: Measure your results

Before finalizing a goal, specify how you'll determine if it's been achieved. You'll also want to evaluate the success of any tactics used, so you'll know which ones to repeat, change, or drop.

Before finalizing a goal, specify how you'll determine if it's been achieved.

Create a comprehensive marketing plan

What is a comprehensive marketing plan?

A *comprehensive marketing plan*[14] analyzes your farm's current situation, looks at how you're marketing your farm, recommends a strategy to achieve your goals, and tells you how to implement the strategy.

Like a speed or small farm marketing plan, a comprehensive marketing plan should be flexible.

Like a speed or small farm marketing plan, a comprehensive marketing plan should be flexible. If a tactic's not working, you'll tweak it or try something new—you don't have to scrap the whole plan. Unless there's a major change in your circumstances, you can adjust any aspect of your plan as needed.

A comprehensive plan may be useful for several years. It usually takes time to put them into action; most of my clients roll out their marketing plans over two or even three years.

When do you need a comprehensive marketing plan?

The larger and more complex your farm is, the more likely you are to benefit from a comprehensive marketing plan. Established farms —especially multi-generation family farms—tend to be highly diversified with a mix of direct-to-consumer, wholesale, and other distribution channels. The marketing tactics that work for each of these channels and target customers may not work for the others.

While creating a comprehensive plan does require some resources up front, it will be easier to run a thriving farm if you have a plan in place. You'll have systems set up so you don't have to think about your marketing strategy during your busiest times. You'll be able to create a more compelling and consistent experience for your customers.

Having a comprehensive plan helps you coordinate tactics, budget, calendar, and messaging. Once you've mapped out your tactics for

14. Most people don't use the phrase "comprehensive marketing plan" to refer to a full-blown marketing plan—they just call it a marketing plan. I'm using it here to contrast it with the speed and small farm marketing plans.

the entire farm in one plan, you may decide to use one tactic to achieve multiple goals. For example, if you create videos to position your farm as a foodie paradise, you can also use the videos to persuade chefs to buy your produce and boost farmers market sales.

Should you DIY or hire a marketing professional?

Most farms should hire a marketing professional to create a comprehensive plan. You'll get expertise and a fresh perspective. If you're applying for a loan or other funding, you may even be required to have a professional marketing plan prepared by an outside source.

It can cost $2,500 to $5,000 or more to hire a marketing agency or consultant for a comprehensive analysis, strategy, and plan. Contact agricultural extension services, state agriculture departments, foundations, and nonprofits for help. Some write marketing plans themselves, while others will provide funds to hire a professional.

If you have some marketing expertise, you can create your own comprehensive plan. You'll save money, but it can take 20–100 hours to create a full marketing plan. Would you rather spend your time farming or marketing?

Elements of a comprehensive plan

Like business plans, marketing plans don't follow a strict formula—everyone does them a bit differently. Some people create a list of tasks and call it their marketing plan, while others go through a detailed analysis, develop a strategy, and then map out the specifics of who needs to do what and when.

It won't surprise you to hear that I'm in favor of the more detailed approach! If you don't do any analysis, you're likely to try random tactics to see what sticks. Even putting a small amount of time into the analysis will help you to choose more effective tactics.

Executive summary

Create a succinct summary describing your farm, goals, and plan. If you're going to share your plan with funders, employees, consultants, or others, the summary is essential—it's all that most people will read. When someone does read the full plan, the summary sets

Comprehensive Marketing Plan Outline

Executive Summary (optional)

Introduction (optional)

Goals

Analysis

- Market analysis
- Customer analysis
- Competitive analysis
- Branding and messaging analysis
- Communications and promotions analysis

Marketing strategies and tactics

Implementation

the context. Executive summaries are typically written when the rest of the document is complete.

If you're the only one who will ever see your plan, the executive summary is optional.

Introduction

The introduction provides a quick overview of your farm and key marketing challenges. If you're the only one who's using your plan, this can be skipped—though it's a useful way to lay out the basic information.

Goals

Your goals should focus on the success of your overall business, rather than intermediate marketing goals such as getting more Facebook Likes.

Setting clear, actionable, measurable goals is key to creating an effective marketing plan. Your goals should focus on the success of your overall business, rather than intermediate marketing goals such as getting more Facebook Likes.

Here are some good examples of goals for a farm marketing plan:

- Expand CSA revenue by 10% annually
- Increase farmstand profitability by 3% annually
- Develop a new revenue source that can generate $100,000+ in additional sales annually

Analysis

The analysis can have as many sections as you find useful. Here are some typical components.

A market analysis covers topics such as market size (in estimated dollars), pricing, consumer trends, and projected sales. This analysis helps you create a more informed plan. If your farm isn't planning a major expansion and you're not applying for a loan, you can skip this section.

A customer analysis describes the customers you currently serve and your target customers. If you serve different types of customers, your plan should analyze each segment because you may need different tactics and communications channels for each one.

A competitive analysis helps you understand who your peers/competitors are, what they offer, and how your farm is similar or different. It helps shape how you want to be seen in comparison to others. It also helps you identify market opportunities and risks. If you operate in multiple sales channels, your plan should analyze your competition in each market.

A branding and messaging analysis looks at how your farm's brand is expressed in-person, online, and in print. It will evaluate whether your logo, key messages, signage, and other brand collateral convey a consistent picture of your farm.

A communications and promotions analysis will review how you're currently spending time and/or money on communications and promotional tactics. It will include any available performance metrics.

Marketing strategies and tactics

This section lays out how you'll achieve your goals. A marketing strategy is an approach to achieving your goal. You may have multiple marketing strategies for each goal. Each strategy will include one or more tactics. For example, if your goal is to increase CSA revenue by 10% each year, you might develop two strategies, one focused on recruiting new customers and the other on increasing your customer retention rate.

A marketing strategy is an approach to achieving your goal. You may have multiple marketing strategies for each goal.

Implementation

This is your road map for how the work will get done. Some people include an implementation plan in their comprehensive marketing plan, while others create a standalone implementation plan that can be easily updated.

The implementation section of a comprehensive marketing plan should cover budget, calendar, responsibilities, tools and systems, and measurement.

Create a marketing implementation plan

A marketing implementation plan compiles your strategies and tactics for all your goals in one place so you can use it, update it, share it, and stay on top of it.

An implementation plan has five components:

- Budget
- Calendar: Ongoing tasks and project tasks
- Responsibilities
- Tools and systems
- Measurement

Where to focus

MICROFARM MUSTS
Invest in signage or creating an online presence that makes your farm easy to find.

SMALL FARM SMARTS
Use your marketing budget to attract and keep loyal customers.

MIDSIZE FARM MAYBES
Put resources into creating memorable experiences for your customers.

Budget

One of the most frequent questions farmers ask is, "How much should I spend on marketing?" The answer depends on whether you're a micro-, small, or midsize farm.

Microfarms can get by with minimal spending on marketing, since most of your contact with customers is likely to be in-person. You'll need signage for your farm and/or farmers market booth. I also recommend that you create a free one-page website or Facebook page. This is likely to amount to $300–$1000 per year.

Small farms must invest in marketing if you want to scale up. This involves a professional logo, a 4-page (or larger) website, email, social media, and print materials. You may also need to upgrade your signage, product information cards, labels, displays, and infrastructure. While many marketing tools are free or low-cost, they require time and expertise to use effectively, so you may need to budget labor expenses if you're not doing the work yourself. Small farms should plan to spend at least $1,000–$5,000 annually on marketing, and up to $20,000–$30,000 if you're growing fast.

Midsize farms should plan to spend a minimum of 5% of your revenues on marketing (including hiring marketing staff or consultants). If you're launching a new venture or expanding, you may need to spend at least 10–20% of revenues.

Calendar

The task calendar should include all your marketing activities over the course of a year. If you can't plan that far ahead, map out your tasks for three months at a time.

Create two sections—a monthly list for one-off tasks and a list of recurring tasks.

One-off tasks are usually part of a process or project. Some tasks can be done at any time, while others must be completed at a specific time.

Ongoing tasks recur on daily, weekly, monthly, or other regular intervals. Marketing is full of ongoing tasks—social media posts, emails, photos and videos, advertisements, and measurement. Be aware of, and realistic about, how much time these tasks take. Do fewer things and set aside time to measure their impact on your business. Use digital tools to schedule recurring time blocks in your calendar for these regular tasks.

Take a look at the big picture and do a reality check. Can we really accomplish this much? Do we need to scale down our plan or spread it out over a longer period? Can I delegate or outsource tasks to someone else?

Sample List of One-off Marketing Tasks

WHAT	EST. COST	EST. HOURS	WHO	START	END
Review and update our key messages for year	None	2 hours	Co-owners	January	February
Create editorial calendar for year	None	2 hours	Co-owners	January	February
Plan events for year	None	2 hours	Co-owners	January	February
Create a system for organizing or finding photos; move all photos into a single online library	Free-to-low-cost system; labor costs for employee to file images	2 hours to choose system; 10 hours to file images	Co-owners to choose system; employee to file images	February	March

How to create a marketing calendar

1. **Make a list of your recurring tasks.** For example:

- Post on Instagram twice a week at x time of day
- Check Facebook daily at 7am, noon, and 5pm to respond to comments, questions, and reviews
- Email CSA subscribers weekly from the first Thursday in June through the Thursday before Thanksgiving
- Review website analytics on the first Monday of each month
- Contact inactive email subscribers on the first Monday in February and August

2. **Break down the steps for each task.** For example, if your task is to email CSA subscribers every Thursday, here's a list of steps:

- Monday 10 am: Choose topics, write content, create subject line and teaser copy
- Tuesday 10am: Take photos (or use existing photos or find stock photos)
- Wednesday 10am: Draft email, send test email, proofread, and check links, and schedule email to send the next day.

3. **Look for task overlaps.** For example, if you create content for social media, you can reuse the text and images in your weekly email.

4. **Estimate the time** it will take to accomplish each task.

5. **Enter recurring tasks** into your calendar.

6. **Make a list of the one-off tasks** to be accomplished, such as getting your logo designed (or updated), creating key messages, printing signs, and setting up social media accounts.

- List the steps involved for each task
- Estimate how much time each task or step will take.
- Review each step to find dependencies—actions that can't be accomplished until another step has been completed. (For example, you'll need a logo before you can print postcards.)

Responsibilities

Who will be responsible for each task? How much will you do yourself? Can you rely on family members or employees to do some of the work? If your employees don't have the right skills, train them. Think long-term.

Marketing requires knowledge and judgment. Your nephew may have grown up using social media, but there's a difference between posting to express your personality versus maintaining a social media presence for business purposes.

You can also save money by doing your own design and photography. Do you have the time and skill to do these well or would you better off finding someone else for these tasks?

If you delegate tasks, decide whether you or anyone else will need to approve their work before it's published or used. If so, build this task into your timeline.

Tools and systems

Most farms have tools and systems in place that can help with marketing, such as a POS, sales payment app, or an email program. In my experience, few farms are taking advantage of the available technology to help them streamline their marketing efforts. Many technology platforms that were initially developed for one purpose—such as web search, email service, or social media—can now be used for other purposes, such as creating free websites, maintaining a customer database, placing digital ads, tracking inventory, and ecommerce.

Here are some processes that will save you time:

Create an annual editorial calendar so you know which products or services to feature on a specific day of the week (e.g. Meatless Monday) or month of the year (Mother's Day specials every May).

Use image recognition to search your digital photos so you can easily find a zucchini photo when you need one.

Store your logo, brand colors, and typefaces in an app like Canva so you can instantly add them to your designs.

Measurement

Measurement refers to how you'll track, measure, and analyze the results of your marketing plan. There are ways to measure everything if you plan ahead. For example, you can add QR codes to your print materials so you can track where people picked up your postcards.

For detailed recommendations on measurment, see the **MEASURE** section.

DO

"It always seems impossible, until it's done." NELSON MANDELA

TOPICS

- Build a strong brand
- Sell the sizzle
- Build customer loyalty
- Increase sales
- Find new direct customers
- Find and keep wholesale customers
- Grow your website
- Grow your social media
- Get found online

What you'll find in DO

Where to focus

MICROFARM MUSTS

You only need a handful of elements to create the foundation for your brand. Read *What is branding, The 3 Cs of branding,* and *The brand elements.* Create the basic brand elements recommended for all farms.

SMALL FARM SMARTS

You need brand elements that create a distinctive feel for your farm. Follow the recommendations for microfarms, then develop the additional tools recommended for small farms.

MIDSIZE FARM MAYBES

Make sure your employees reflect your farm's brand in everything they say and do. Add the additional elements recommended for midsize farms.

Tactics to achieve specific goals:

If you're more of a doer than a planner, this part of the book is for you! Each section provides step-by-step guidance for achieving a specific goal for your farm.

Build a strong brand. Your farm's brand is an intangible asset that helps you attract your target customers. Learn the essential brand elements that your farm needs.

Sell the sizzle. People don't buy products, we buy feelings. Learn how to leverage sight, sound, smell, taste, and touch to associate positive emotions with your farm.

Build customer loyalty. A loyal customer base is essential to a thriving farm. Learn how to build loyalty by appreciating your customers' relationship with your farm.

Increase sales. Get ideas for tactics that can increase sales using your current offerings and distribution outlets.

Find new direct customers. Learn how to develop a strategic approach to attracting new customers.

Find and keep wholesale customers. The secret to wholesale success is not just getting a new account, but keeping it. Learn how to get in and stay in.

Grow your website. Guidance for how to create a one-page, four-page, and full website.

Grow your social media. Guidance for creating a passive or active social media presence.

Get found online. The right tactics for getting found online depend on your farm's size and goals. Learn the essentials for your farm.

Build a strong brand

What is branding?

What comes to mind when you see each of the images to the right?

If you're like most people, you probably had different associations when you saw each cup. Dunkin' and Starbucks both sell coffee! So why does each brand generate a unique response?

A brand is a collection of tangible and intangible characteristics that people associate with a business or its offerings. As soon as we become aware of a business or product, we start forming unconscious associations. It may take more than one encounter, but sooner or later, a brand lodges in our brains. Our knowledge, beliefs, and feelings about the business grow with each contact.

It's hard to differentiate your farm based solely on your offerings, production methods, or other tangible characteristics, because customers may not be able to perceive how you're different from the other farm down the road. The intangible aspects of your farm that are part of your brand can help you stand out.

Starbucks and Dunkin' coffee cups

(Photo : Michelle Johnson)

The 3 Cs of Branding: Clear, Compelling, Consistent

Your brand consists of many different elements. How can you fit all the pieces together into a coherent identity? Make sure your brand is *clear, compelling,* and *consistent.*

Clear. Your target customers "get" who you are at a glance. You emphasize a few key messages that you want people to know about your farm.

Compelling. You use words, images, and concepts that are true to your purpose and likely to resonate with your target customers. You employ stories, emotions, and the five senses to help customers feel connected to your farm.

Consistent. Your words, images, and actions reflect your brand values. You use the same language, color palette, images, and tone of voice for all your communications.

Brand Elements

FOR ALL FARMS

- Farm purpose and values
- Farm name
- Farm personality
- Logo
- Typefaces
- Color palette
- Signage and displays

ADD > SMALL + MIDSIZE FARMS

- Farm statement
- Descriptor and/or tagline

ADD > MIDSIZE FARMS

- Tone of voice
- Brand guide
- Brand training tool

It takes work to find the right words and images to express who you are, what you do, and why and how you do it, but once you've developed these elements, you'll use them over and over. You can't completely control how customers perceive your brand, because their worldview and life experiences may color their impressions. If your brand is coherent, however, your customers' perceptions of your farm are more likely to match your true identity.

The brand elements

Brand elements are the building blocks you'll use to create a strong brand. The basic elements are essential for all farms—regardless of size—because they're the foundation for the subsequent components you'll need.

As your farm grows and diversifies, you'll get a clearer sense of who you are and how you want to be seen. You'll incorporate additional elements as needed.

Basic brand elements for all farms

Farm purpose and values

Your farm purpose is the reason why you farm. It's the cornerstone of your brand identity and drives your business, operating, and marketing decisions. Having a clear purpose inspires your employees, attracts new customers, and builds customer loyalty.

Whether you're farming because it's part of your cultural heritage, you're running a business, you're on a mission, or you feel compelled to work the land, your motivations are the heart of your brand.

Your farm values are the guiding principles for how you (and your farm) view, operate, and interact with the world. Your values may focus on your *beliefs* (the why) and/or your *operating principles* (the how).

You can choose to share your purpose and values on your website and social media profiles or just use them to inform the other brand elements that you make public.

How to clarify your purpose and values

1. Write down your answers to the following questions:

- Why did I start (or join) this farm? Why do I continue, despite the challenges?
- What guiding principles does my farm follow when interacting with family, employees, the land, and the animals?
- What guiding principles does my farm follow when interacting with customers, peers/competitors, suppliers, and the wider community?

2. Review your writing. What stands out? Circle any words or phrases that resonate with you.

3. State your farm's purpose and values in a phrase, sentence, list, or paragraph (100 words maximum).

Aishah Lurry, owner of Patagonia Flower Farm in Patagonia, AZ. (Photo: Patagonia Flower Farm)

Patagonia Flower Farm expresses its purpose and values

"The foundation of a healthy micro-farm starts with biologically diverse fields and healthy soil. Operated since 2017, Patagonia Flower Farm is your local source for fresh flowers that are never imported. Our organic practices are as healthy for the Earth as they are for your home and office."
From the website's About page

Farm name

Your farm name is how people refer to and remember you. While having a meaningful and memorable name is ideal, it's not necessary. Customers choose farms based on offerings, convenience, growing practices, and purpose, not because they have a fabulous name.

Your farm name conveys a subtle message about you, your values, and your personality. Your name can be descriptive, tell a story, reference your purpose, promote a sense of place, or even make people laugh.

- Many farms are named after their owners, their location, what they grow or produce, and/or how they farm
- Some names mention a specific feature (a hill, a river, or crossroad) that resonates with people who are familiar with your location
- Other names are fanciful (Blue Sky Farm) or funny (Fool's Gold Farm)

How to find out if a name is already in use

Here's my common-sense advice on how to vet a potential name for your farm or other aspect of your business.[15] If you use a name that's trademarked by another business, you could get a *cease and desist* letter demanding that you give it up. So find out if the name you want is already in use.

Enter your proposed name into an internet browser. If you're considering a common name like "red barn farm," you'll find multiple farms with that name. Because the name and the words it contains are generic, it's probably safe to use it (provided that your farm is operating in a different area).

Visit a domain registrar (such as domain.com or godaddy.com). Type in a URL that incorporates your potential farm name (e.g., redbarnfarm.com or redbarnfarm.farm). If the URL you want is available, that's a good sign. URLs don't always match farm names, so it's not a guarantee that the name isn't in use.

Check to see if the name is trademarked. Visit the US Patent and Trademark Office website at https://www.uspto.gov/trademarks/search, click "Search our trademark database" and choose "basic word mark search." If no one owns an active trademark for the name, you're probably okay.

Farm personality

Brands have personalities, just like people. Think about your best friend. What three words would you use to describe them?

- If they talk to strangers on the street, you might say *friendly*
- If they're always wearing the latest clothes, you might say *trendy*
- If they like to suggest new things to do, you might say *fun*

The brand personality your farm projects can help determine the type of customers you attract. If customers enjoy your personality, they'll be drawn to your farm, just as we're drawn to certain types of people. You can't fake your personality; customers will sense if you're pretending to be something you're not.

15. NOTE: I'm not an attorney and my comments are not legally binding. If you want to make absolutely sure that your proposed name or tagline is safe to use, hire an intellectual property attorney.

Here are some examples of brand personality traits and behaviors that might reflect your farm:

Friendly. Your farm feels welcoming. You and your employees smile at customers, know the regulars by name, and start conversations.

Family-oriented. You offer products and activities that appeal to kids, have kid-sized equipment, give kids free snacks, and understand their parents' needs.

Community-oriented. You're locally-owned, hire community members, use local vendors, seek community input, donate to the community, and hold open houses.

Knowledgeable. You hold cooking demonstrations, share recipes, and provide information about how to use products. Staff offer help when needed.

Edgy. You use bold colors, play the latest music in the farm store, and showcase new products, services, and experiences.

Sophisticated. You provide high quality, unusual offerings. Your signage and packaging are well-designed. You enjoy sharing your history and culture.

Open-minded. You try to understand, appreciate, and welcome people of all cultures and worldviews. People of all races, ethnicities, ages, and backgrounds feel comfortable at your farm.

How to clarify your brand personality

Brainstorm a list of characteristics that you believe your farm reflects.

Ask a few loyal customers to describe your farm personality in 3–5 words or phrases. Do any of their words match the characteristics that you listed?

Select 3–5 adjectives that describe how you want your farm's personality to be perceived.

Note: Your brand personality is a tool that will help you choose language, colors, images, and offerings that resonate. It's only for internal use!

Where to focus

MICROFARM MUSTS

If you need a logo in a hurry or are unable to hire a graphic designer, you can use free or inexpensive software, such as Canva or Hatchful, to create your own.

SMALL FARM SMARTS

If you've outgrown your first logo and are still on a tight budget, you can get an inexpensive logo design using a freelance service like Fivver or 99designs. Prices depends on your choice of designer and how many options you want; $50–$200 is a reasonable range for a simple image.

MIDSIZE FARM MAYBES

Upgrading your logo can increase brand visibility and help differentiate your farm. Invest in a designer who can update your image yet express your new identity. Expect to pay at least $250 for a very simple design and $1,000 or more if it involves illustration or more intricate work. Agencies that specialize in branding will charge more; one of my clients paid $10,000 for a design from a top agency.

Your farm logo, typefaces, and color palette

A logo is a visual symbol that represents your farm. Your logo is not your brand; it's a symbol that represents your brand.

Once a customer recognizes your logo, it automatically evokes associations about your farm. Your logo should reflect your brand personality and values, so it creates the perceptions you want people to have about your brand.

When you consistently place your logo on your signage, cards, labels, and website, it helps people recognize your farm, makes you look professional, and establishes trust. People know who you are and that you stand behind your products, services, and experiences.

Logos are often the starting point for your entire visual presence. The design of your website, signage, and other materials should incorporate the look and feel of your logo. When you create your logo, try to set the mood or tone that you want people to associate with your farm. If you want them to see your farm as family friendly, use warm colors and a welcoming typeface. If you're growing flowers for high-end weddings, use cooler colors and a more sophisticated tone.

A logo typically consists of 1–3 elements:

Your farm or product name

Most logos include the name of the farm, with the letters rendered in a specific shape and typeface. Typefaces have strong personalities, so choose one that matches how you want people to think about your brand. Review your brand personality traits and look for a typeface that fits.

Color(s)

Most business logos choose two or more colors to represent their brand personality and the feelings they want to evoke. Think you're not using color because your logo consists of black text on a white background? Wrong! Your colors are black and white. Viewers perceive the absence of color (white space) as a color. It's an aesthetic choice that sets the tone for your brand.

How to prepare your logo for multiple uses

Your logo will be used in many ways—print materials, signage, website, social media. You, or your designer, should create variations of your logo in different shapes and formats so it will be available when you need it.

Here are some variations that may be useful:

- a version that looks good on a website, social media, or other digital channel (72 dpi in .png format)
- a high-resolution version for print (300 dpi in .eps format)
- a rectangular version that looks good on a website or email header
- a square or round version that looks good at a very small size (for social media profiles)
- a version that looks good in black-and-white
- a version that incorporates your descriptor and/or slogan
- a version that just contains your farm name in your signature typeface

The colors in your logo will typically be used as the main colors in your signage, packaging, and other visual elements. You may want to choose a few additional colors or shades that are compatible with your main colors to use as secondary colors, so you have a broader color palette to work with. If you're working with a graphic designer, they can provide you with a chart of your main and secondary colors.

A symbol

A brand symbol can be an illustration, a photo, a squiggly line, or any visual element. The symbol may or may not incorporate the company name or product into the image. Some brand symbols can become so iconic (hello, McDonald's Golden Arches) that their company name isn't as essential.

Some farms use a simple image (a flower, a barn, an apple) in their logo to express their offerings; others depict the location that their farm's named for (a rising sun, a creek, a hill).

While you don't have to use a symbol in your logo, it will come in handy. For example, logos often need to be shrunk down to small sizes. If necessary, you can use the symbol to represent your brand when your farm name would be too tiny to be legible.

Photo: Myrna Greenfield

Signage, displays, labels, packaging, and other visual elements

If you welcome visitors to your farm or sell at farmers markets, you'll need signage and displays. You may also need packages, product labels, or tools to present your farm and offerings.

These visual elements should reflect the core aspects of your brand—your purpose, values, name, logo, typefaces, and color palette. When you present a consistent image, you come across as a familiar, trustworthy friend.

The seafood banner in the photo to the left was the only eye-catching signage I saw at a farmers market booth; I bought seafood there for weeks before I discovered that the business actually had a name. If they change their location, I might never find them again.

I also saw a farmers market booth with a nice banner, but it was so hidden by baskets and buckets that I couldn't see the name of the farm. Your signage helps customers identify and remember you. Make sure we can see it!

Some farms create generic-looking labels for their value-added products that don't reflect their brand. Your packaged products sit on your customers' tables or in their fridges. They reflect your brand; every time customers see your products, it will remind them of your farm.

Your print and digital materials or displays don't have to be fancy or expensive. They should be clear (quickly convey your message), compelling (appeal to your target customers), and consistent (they look like they come from you).

Additional elements for small and midsize farms

As a small or midsize farm, you probably sell to multiple channels, have other people working with you, and are building your customer base. You'll need all the brand elements previously described, plus the ones we'll discuss next.

Screenshot of the top portion of the home page for Clean Greens Farm & Market.

Create a "farm statement"

A "farm statement" can be a phrase, a few words or bullet points, or 2–3 short sentences that state or imply how your farm satisfies your target customers' needs and desires. In marketing jargon, this is known as your *value proposition*.

Your statement should say what you offer and convey how it's meaningful or special for the type of customers that you want to attract. An effective statement will prompt your target customers to think or say, "Wow, that's my kind of farm! Tell me more."

Let's look at two examples.

Clean Greens Farm & Market (https://www.cleangreensfarmmarket.com/) is a nonprofit urban farm in Seattle, WA. Their offerings and mission—"Making organic produce widely available and easily accessible to all people in the community"—are conveyed through their name, logo, text and images on their website.

Screenshot of the mobile-responsive home page for Owl's Nest Farm.

Owl's Nest Farm (https://www.owls-nest-farm.com/) in Upper Marlboro, MD, has a simple statement at the top of their website home page that is complemented by the homey image of colorful vegetables set on top of a tablecloth.

They go on to state what they offer in a brief paragraph that expresses their brand personality and values.

How to write your farm statement

1. **Write down your answers** to the following questions:
- What types of products, services, and experiences does our farm offer?
- What are we known for?
- What needs, values, or interests do our most loyal customers share? What do they love about us? What motivates them to keep coming back?

2. **Review your writing.** Circle any words or phrases that resonate with you.

3. **List the top 3 points** that you want customers to know about your farm.

4. **Draft a sentence for each point.**

5. **Combine your points** into a few words, bullet points, or sentences.

6. **Simplify your statement** until it reads like something you'd actually say.

Descriptors and taglines

Descriptors and taglines consist of a few words or a short phrase —eight words or less—that help your target customers understand and identify with your farm. Descriptors and taglines can be used alongside your logo or included when appropriate.

A *descriptor* describes your farm. It's useful if your farm is named after your family or location and people can't tell what type of farm you are from the name.

A *tagline* inspires people to feel connected to your farm's purpose, values, or personality.[16] An effective tagline evokes a feeling—joy, laughter, inspiration—that you want people to associate with your business. Taglines are typically written using the root form of a verb, such as "Shop," "Visit," or "Eat." It's a command that directs the customer to think, feel or do something.

---

16. Taglines are also called slogans. While slogans are more often used in advertising campaigns, rather than for overall branding, most of us use the terms interchangeably.

Comparison of Descriptor and Tagline

DESCRIPTOR	TAGLINE
Describes something about what type of farm you are or what you offer	Conveys a feeling or values
Uses descriptive words	Uses inspirational words
Literal	Figurative

Here are a few classic taglines:

- Apple: *Think different*
- Nike: *Just do it*
- Burger King: *Have it your way*

None of these phrases tell us anything about what the business offers. They invite us to think, feel, or act a certain way.

Sometimes, a single, well-worded phrase can function as *both* a descriptor and a tagline. For example, Green Mountain Girls Farm in Northfield, Vermont uses one phrase to both describe and inspire.

GREEN MOUNTAIN GIRLS FARM
EAT. STAY. FARM.

Green Mountain Girls is a diversified vegetable and meat farm that offers farms stays and other agritourism experiences. The phrase, "Eat. Stay. Farm." succinctly conveys what they offer, so it works as a descriptor. Each word is a command designed to appeal to people who want to spend time on a farm, so it also functions as a tagline.

It's usually easy to create a descriptor and hard to write an inspiring tagline. Some farms have one or the other, some have both, and some have neither. Descriptors and taglines are useful, but not essential. If you can't find words that feel right for your farm, let it go for now and try again when you're ready.

Brand elements for midsize farms

You may already have most of the brand elements in place. You probably divide your marketing tasks among several people or work with outside consultants or agencies. Your biggest challenges are consistency and efficiency. These additional elements can help your farm save time while projecting a more consistent brand image.

Tone of voice

The *tone of voice* your farm uses in all your communications should reflect your farm's values and personality, just as your logo, color palette, typefaces and other visual elements do. A tone of voice can be friendly or formal, laid-back or excited—it can change the meaning of your words from a suggestion to a command.

Think of the difference between a text message and a business letter. A text is short, friendly, and casual with plenty of abbreviations, emojis, and exclamation marks. You expect your friends to text like this. If you get a business letter from your insurance company written the same way, you might not trust them with your policy.

People will detect your tone of voice from your website, social media posts, signage, and verbal communications. For example, if your orchard is plastered with "No entry" signs on rows of apples that aren't ready to pick, you'll come across as nagging; if you tie colorful string and post a funny sign to block entry to the same row, the message will come across in a more playful, friendly manner.

Signage at Red Fire Farm in Granby, MA. (Photo: Red Fire Farm)

Projecting tone of voice in signage

Red Fire Farm in Granby, MA, gets their tone of voice in their signage just right. Their bright yellow signs are hard to miss. The fanciful script is artistic, friendly, and informal. Their tone is direct, not bossy.

How to create guidelines for tone of voice

Specify the tone of voice you want to convey in your communications, using plenty of examples. This will make it easier for anyone who creates text or images for your farm to convey a consistent tone.

Here are some typical tone of voice examples that you may want to include in your guidelines:

- Are contractions acceptable or do you want almost all words to be written out? For example, do you want people to write "Don't" or "Do not"?
- Are emojis acceptable? If so, are there any types of emojis that are inappropriate? Do you want them to be used sparingly or are multiple emojis acceptable?
- Are slang words acceptable? What about swear words?
- Are technical terms, such as "Integrated Pest Management" acceptable? If so, should they be spelled out and defined before you refer to them in abbreviated form (e.g., IPM)?
- Is it okay to use an exclamation mark after every sentence or do you want them to be used sparingly?

Don't have time to create full guidelines? Give anyone who works on your marketing a few general principles, such as "Use a friendly tone, but keep it businesslike."

When everyone involved uses the brand guide and tone of voice guidelines, all your farm communications will look and sound like they're coming from one person.

Brand guide

A *brand guide* is a document that includes all the brand elements that you use on your farm. Upload the brand guide and your brand elements into an online folder for easy access. Trust me, you'll save time when all your brand elements are stored together!

Creating a consistent look and feel for your marketing is easy when only one person is doing the work. As you delegate tasks to employees or consultants, it's harder to maintain that consistency. When everyone involved uses the brand guide and tone of voice guidelines, all your farm communications will look and sound like they're coming from one person.

Typical elements for a brand guide

- Purpose and values
- Farm name (and any names you use for your products, services, or other aspects of your farm)
- Personality
- Farm statement
- Descriptor and/or tagline
- Logo (and variations)
- Typefaces (typefaces, sizes, and spacing)
- Color palette (what main and secondary colors to use)
- Tone of voice

Brand training tool

Everyone who works with your farm must understand your brand purpose, values, and personality and be able to express them through their words and actions.

Easier said than done. Branding is a hard subject to convey because it deals with so many intangibles. The brand guide described above is a wonderful tool for anyone who's doing your marketing, but most of your staff won't be interested in reading a longish, somewhat technical document.

How to create a brand training tool

Be creative! Here are a few ideas:

Develop a story or a phrase that sums up your brand in a few words. Your brand statement may encapsulate everything you need to say.

Create a one-page document that highlights a few key brand elements, such as your farm's purpose, values, and personality.

Make a video. Ask someone to interview you on their smart phone. The video is for internal use, so it doesn't need to be professional quality. Once you have the video, you can incorporate it into training for new employees.

Write a comic strip. A comic strip is a series of images that tell a story. If you or someone on your farm is artistic, you may be able to convey what your brand is about in a small, fun little book. Stick figures can work just fine, but if no one likes to draw, you could find some free photos online to illustrate your key points.

Give your employees a brand training tool that makes your brand easy to absorb, remember, and use.

Let your brand evolve

If you've gone through all or most of the exercises in this chapter, give yourself a pat on the back! It takes self-awareness, patience, and persistence to develop a clear, compelling, and consistent brand.

Being consistent doesn't mean that you have to keep using the exact same words and images over and over. As long as everything you say or do projects the same values and personality, you can vary your content to create different social media posts, ads, and offers that target specific segments of your customer base.

Building a brand is an ongoing process. You'll develop new insights as you run your farm. Don't let my call for consistency stifle your creativity. You can always find new and better ways to express the essence of your farm.

Sell the sizzle

It takes more than heirloom vegetables, fluffy wool, or chunky salsa to make people fall in love with your farm. You need to build an emotional connection with your customers.

Words are important, but your non-verbal communications hold more sway. Positive sensory experiences, fond memories, and natural endorphins are customer magnets.

The phrase, "Don't sell the steak, sell the sizzle" was coined a century ago by Elmer Wheeler, an ace salesperson and marketer who became known as "Mr. Sizzle." He saw that people got hooked by the senses and emotions that an object arouses, not just the object itself. Hearing the sizzle of the steak in a pan, smelling that steak cooking, or even seeing a photo of a browning steak on a grill will make a steak lover crave it.

We don't buy products, we buy feelings

Human beings like to think we're rational creatures, ruled by that great big brain of ours. But when it comes to most purchasing decisions, we don't buy products or services, we buy feelings.

When it comes to most purchasing decisions, we don't buy products or services, we buy feelings.

Making even small decisions takes up a lot of our brain's processing time. If every purchase involved a detailed decision-making process, we'd never get anything done. We make most decisions through our hearts, not our brains.

In fact, "95% of our purchasing decisions are driven by unconscious urges, particularly our emotions," according to Harvard Business School professor emeritus, Gerald Zaltman.[17]

Emotions can process sensory input five times faster than the conscious, rational brain. Our feelings are triggered before our thoughts.[18]

17. Mahoney, M. (2003, January 13). The subconscious mind of the consumers and how to reach it. *Harvard Business Review.* https://hbswk.hbs.edu/item/the-subconscious-mind-of-the-consumer-and-how-to-reach-it

18. Hill, D. (2007). *Emotionomics: Leveraging Emotions for Business Success.* KoganPage.

Images, sounds, tastes, touches, and smells that evoke emotions, memories, and associations can have a powerful impact on the products or services we choose, social media pages we follow, and brands we love. When you associate your farm with emotions that appeal to your target customers' inner needs and desires, they're more likely to feel an affinity with your brand.

When you associate your farm with emotions that appeal to your target customers, they're more likely to feel an affinity with your brand.

Leveraging emotions is a powerful tool that should be used in an ethical manner. If you manipulate someone into buying a product or service they don't want or need, they might resent you later. When you provide a customer with a useful product or service that gives them a sense of belonging, purpose, or inspiration, they're more likely to become a loyal customer.

Using the five senses to generate emotions

Sight

Images

A brand without images is like a song lyric without the music; you're not getting the full effect. Look for images that tell a story or rouse an emotion. Choose images that match your farm's values and personality.

Video

Video can be more potent than still images. It captures our attention and compels us to want to see what comes next. It appeals to our ears as well as our eyes. Social media ads, stories, and posts that contain videos or carousel (rotating) images attract more engagement than those with still images.

Don't be shy about using video whenever you can. Viewers are looking for authenticity from your farm, not slick production values. Live or unedited videos can be perceived as more trustworthy than prerecorded productions.

Colors trigger feelings faster than words.

Color

Colors trigger feelings faster than words. According to studies, up to 90% of a person's first impression of a product may be based solely on your use of color.[19]

Marketing guru Seth Godin used the concept of a *purple cow*[20] to make the case that you need to be remarkable in order to stand out in a world awash with marketing messages. An ordinary cow is unlikely to get noticed; a purple cow would certainly attract attention!

Certain colors tend to generate specific associations or emotions. For example, red is associated with passion, energy, activity, and stimulating the appetite; green with calm, nature, growth, and prosperity; and blue with the sea, sky, peace, harmony, and relaxation.

When choosing brand colors, select shades that inspire the desired feelings. Reactions to color aren't universal across cultures or backgrounds. Show your proposed palette to your loyal customers to see how they respond.

Symbols

Symbols are a powerful way to evoke a feeling. Merriam Webster defines a symbol as a "visible sign of something invisible."[21] A symbol can be an image, gesture, object, or idea that represents something other than itself.

There are all kinds of symbols, from pictographs and emojis to flags, hand signs and status symbols. Words are a type of symbol but our brains process images 60,000 times faster than text.[22] Visuals are more likely to trigger emotions, because images and emotions are processed in the same area of your brain.

---

19. Singh, S (2006, July 1). Impact of color on marketing. *Management Decision.* https://www.emerald.com/insight/content/doi/10.1108/00251740610673332/full/html

20. Godin, S (2009). *Purple Cow, New Edition: Transform Your Business by Being Remarkable.* Portfolio.

21. Merriam-Webster. (n.d.) Symbol. In *Merriam-Webster.com* dictionary. Retrieved August 11, 2022, from https://www.merriam-webster.com/dictionary/symbol

22. Humans process visual data better (2014). *Thermopylae Sciences and Technology.* https://www.t-sciences.com/news/humans-process-visual-data-better

A visual symbol can be a shortcut to creating the associations that you want people to have with your brand. Words are often unnecessary; the symbol triggers an unconscious response.

The symbols for various certifications and labels, such as USDA Certified Organic, Fair Trade, or Non-GMO, are tools that can associate your brand with certain values.

Look for popular symbols that you can leverage for your farm. For example, the family farm is such an iconic and beloved American symbol that the image of a farmhouse, field, silo, or barn spontaneously generates warm feelings.

Hearing

Music, the sound of someone's voice, a laugh, a sneeze—sounds grab our attention. We often associate a sound with the experience or setting where we heard it. Hearing that song or sound again triggers our memory of that moment and what we felt at the time.

If you hold concerts at your farm or play background music at your farm store, choose musical genres that fit your brand. If you seek to welcome customers from many different cultures, offer a range of musical styles.

When you make videos of your farm, look for settings where viewers can hear sounds from your farm—cows mooing, tractors roaring, or birds chirping.

Taste

Research shows that people believe that foods that come directly from farms taste better, so they're predisposed to assume that your food will taste good. Offer samples and tastings whenever possible.

When communicating online or in print, use descriptive language and gorgeous photos of food and drink; you can make your customers' mouths water without tasting a thing.

Smell

Our taste buds are alerted by smells. Entice customers with the smell of hot cider donuts frying or fresh lavender growing in your fields.

If you're not selling in person, arouse the sense of smell on your website or social media by choosing images of things that are associated with strong smells (BBQ, a wood fire, garlic).

Touch

The desire to pet a horse, hold an apple to test its firmness, or pick a ripe strawberry is hard to resist. Health and safety regulations may limit your ability to let customers touch everything on your farm, but offer any opportunities you can. PYO, petting zoos, and hands-on workshops offer controlled opportunities for customers to get in touch with your farm.

Other ways to evoke emotions

Humor and physical responses

Humor is a powerful tool for building associations with your brand. Laughter physically relaxes you and lifts your mood. Other physical activities and sensations—a hike through your orchards or eating an ice cream cone at your farmstand on a hot August night—also generate positive body chemistry that triggers good feelings.

Nostalgia, rituals, holidays, vacation, and other stimuli

Appealing to nostalgia, particularly childhood memories, creates warm associations for most of us. Farms typically decorate with old glassware, straw brooms, gourds, and other icons of traditional farm life—evoking a warm response from people who probably didn't grow up on a farm.

Local festivals, holidays, and other annual activities have built-in, usually positive associations. If any of these are a good fit with your farm, look for ways to associate such events with your farm.

Activities and experiences

"A powerful brand experience," former Sephora executive Mary Beth Laughton stated, "is "memorable, shareable, repeatable, and addictive."[23] When customers go out of their way to buy your potatoes,

---

23. Walton, C. (2020). Lulemon's Mirror Acquisition is Next Level Omnichannel Retailing at its Finest. *Forbes*. https://www.forbes.com/sites/christopherwalton/2020/07/08/lululemons-mirror-acquisition-is-next-level-omnichannel-retailing-at-its-finest/#549d75c485c6

they're not just buying spuds, they're buying the experience. As a farmer, you represent a lifestyle that many people crave—connection to nature or animals, growing food with your hard work, the taste of your harvest on the table.

Activities and experiences are excellent ways to make your brand compelling to your target customers. While it's easiest to create a great experience by offering on-farm activities (PYO, workshops, a petting zoo, event rentals), most farms have many opportunities to create positive brand experiences for your customers.

You can evoke a "you-are-there" feeling by offering virtual tours and tastings or online workshops. Use email, social media, texts, and other digital communications to share farm stories and updates. Describe what it feels like to start working at dawn and see the sun rising over your fields or chase an unruly cow.

Build customer loyalty

Marketers estimate that it's 5 to 25 times cheaper to keep an existing customer than to find a new one.[24] While you do need to keep adding new customers (because no matter how much your customers love your farm, you'll have some attrition), always treat your existing customers as the treasures they are.

There are other farms where your customers can find similar products, services, or experiences. Once they've connected with your farm, you must motivate them to keep coming back

Build customer relationships over time

Customer relationships are just like any other type of relationship—if there's mutual interest, it will deepen with time. Someone visiting your farm store for the first time will feel less connected than a customer who's bought your Thanksgiving pies for the last ten years. Your goal for first-time customers is to motivate them to return; you want your long-term customers to feel valued and stay loyal.

Where to focus

MICROFARM MUSTS
Focus on building relationships with your customers. Develop strategies for encouraging first-time customers to return and helping long-time customers feel appreciated.

SMALL FARM SMARTS
Consider starting a customer loyalty program.

MIDSIZE FARM MAYBES
Reach out to each type of customer you serve. Create targeted offers and messages that will appeal to each segment.

24. Gallo, A. (2014, October). The value of keeping the right customers. *Harvard Business Review*. https://hbr.org/2014/10/the-value-of-keeping-the-right-customers

How to detect a first-time customer

Spotting people who are visiting your farm or farmers market booth can be tough, especially when it's busy. If most of your customers are regulars, they're likely to be offended if you don't recognize them.

Look for people who are spending a lot of time looking around or moving hesitantly and say, "Can I help you find anything?" Their response may tell you if they're a newcomer.

You can also ask neutral questions that can apply to new or returning customers, such as, "Are you in our loyalty program?" or "Are you on our mailing list?"

Encourage first-time customers to come back

- **Be helpful**. Ask if they're looking for anything in particular
- **Be hospitable**. Offer a free cup of hot cider or piece of your maple candy
- **Make it easy to visit or shop.** Post wayfaring signage throughout your farm and store so they can find everything they need
- **Share your farm's values and story** via family photos, table tents, posters, or other media
- **Ask for feedback.** When a customer's at the cash register, ask if they found everything they were looking for or how their visit went
- **Incentivize them.** Drop a discount coupon for their next visit into their bag at checkout[25]

While you do need to keep adding new customers . . . always treat your existing customers as the treasures they are.

Recognize customers on their one-year anniversary

Acknowledging that someone has been a customer or CSA member for a year is an optional but fun way to recognize them. If you have a loyalty program, CSA or POS system, you can use it to identify how long someone's been a customer.

A simple thank you to make them feel appreciated is usually sufficient. Send a message like: "Happy anniversary! Thanks for being

25. Put an expiration date on your coupons to encourage them to return while the memory is still fresh.

a customer (or on our email list) for a year." If you want to do more, send a certificate to redeem for a free gift on their next visit.

Appreciate your long-term customers

Acknowledge and appreciate the customers who have been with you for 3, 10, 20, or more years. Give long-term customers free merchandise or a discount on a regular basis; surprise them with a random gift.

Rewards don't have to be monetary; your long-termers may be pleased with a shout-out in your customer emails or on social media. They'll also enjoy having more contact or access than newer customers, such as an invitation to an end-of-season potluck or cider tasting open only to seasoned customers.

Reach out to lapsed customers

If you're able to identify people who haven't visited your farm or opened an email for a certain amount of time,[26] send them a targeted email or text to encourage them to return. Sometimes a simple, "We miss you!" will persuade someone to return, while others will need a discount or other incentive. Even if you only regain a few customers, it's worth the effort.

How to build your email list

Set up systems to collect customer contact information (with their permission). For example, if you hold events that require an online RSVP, include a box to check if they want to be notified about future events or activities. If you sell online, ask customers to opt-in to join your list. Loyalty programs also serve this purpose.

Place a digital tablet or email sign-up form on your counter. Offer a raffle or other incentive to encourage them to join.

Offer to take customer photos at PYO or at events to email to them afterward.

Welcome new email subscribers with a series of automated email messages to introduce them to your farm; this will encourage them to open subsequent emails.

---

26. Due to changes in privacy practices, it's more difficult to verify if someone has opened an email unless they've clicked one of your links.

Be careful how you define a lapsed customer. For example, if you have a PYO farm or run an annual event, you may have some customers you only see once a year. In that case, you might want to wait several years before considering them lapsed.

Develop a customer loyalty program

Customer loyalty programs offer repeat customers a reward. Some customers join loyalty programs because they want a bargain, others because they feel an affinity with your farm.

Some loyalty programs are purely transactional; buy ten dozen cider donuts and get the next dozen free. Others offer unique experiences. For example, if you have a Beer of the Month club, you can invite members to an annual tasting party just for them.

How to protect customer privacy

US data privacy regulations vary by company, industry and state. The regulations apply to where your customers are located, not just to where your farm is based. As a small business, you may be exempt from some requirements, but check the regulations for the states where you do business.

Here are the customer data privacy practices I recommend:[27]

Research how your website, email service provider, loyalty program, social media apps, POS, credit card processor, and other apps collect, store and use customer data.

Use a free or inexpensive *privacy policy generator* to craft a privacy.

Post your privacy policy on your website.

Create a *cookie notice* that informs customers about how you collect data and how to opt out.

Always disclose your farm's privacy practices whenever your customers sign up for a loyalty card, email list, or any system that collects data.

---

27. Consult an attorney to ensure that you're complying with all necessary requirements.

Are loyalty program worth it?

Research shows that loyalty programs *are* effective at cultivating strong customer relationships, but don't usually get customers to visit more frequently or spend more per visit. A 2021 study found that consumer loyalty programs increased the value of a customer by nearly 30% over a 5-year period.[28]

You can lose money from your loyalty program if your overhead costs are too high or benefits too generous, so weigh the total costs versus benefits. Did sales increase after you instituted the program? Are you getting more repeat customers? Are you getting more online reviews raving about your farm?

You can lose money from your loyalty program if your overhead costs are too high or benefits too generous, so weigh the total costs versus benefits.

Measure qualitative responses, too. Do your loyalty customers say anything positive about the program?

DIY vs. commercial loyalty programs

DIY

Skip the monthly fees, customize your program, and maintain control over your data by creating your own loyalty program. You'll have to do all the set-up, bookkeeping and administrative work yourself. Keep it simple—create a paper or plastic card that gets punched, stamped or swiped each time a customer spends a certain amount.

You can offer meaningful loyalty benefits that require minimal administration. For example, you can hold an annual harvest party to thank your CSA members. You already have a list of CSA members to invite, so there's no any extra paperwork to figure out who qualifies.

Commercial loyalty programs

Commercial loyalty software or apps can be stand-alone programs or add-ons to your POS system, website, or other service. Commercial programs use a customer name, phone number, or credit card

28. Gopalakrishnan, A., Zhenling, J., Nevskaya, Y., Thomadsen, R. (2021). Can Non-tiered Customer Loyalty Programs Be Profitable? *Marketing Science*. https://doi.org/10.1287/mksc.2020.1268

to track purchases and dispense rewards. Rewards can be offered based on purchase frequency or amount. Most commercial programs enable you to personalize offers and messages to target your customers' needs and interests.

Expect to pay a monthly subscription fee—based on how many customers participate—and various transaction fees. For example, at the time of this writing, Square charges $45 a month per location for 500 points. You'll use one point each time a customer enrolls, makes a purchase, or redeems a loyalty award; you'll pay additional fees if you want to email or text customers through Square.

Reach out to each type of customer

Sending customers targeted messages and offers that reflect their interests, needs, and relationship with your farm helps build customer loyalty.

Sending customers targeted messages and offers that reflect their interests, needs, and relationship with your farm helps build customer loyalty. Groups of customers with shared characteristics are called *customer segments*. When you create relevant segments, you can choose marketing tactics that are most likely to appeal to people in that group.

This section describes different ways to segment customers. Choose the segments that you're able to identify using available data and that will help you achieve your goals.[29]

By products, services, or experiences used

People who buy the same products, services, or experiences often share similar interests, habits, and needs. Code or tag the customers in your database by products, services, or experiences used (CSA members, PYO customers, and holiday pie customers), so you can target them with the right tactics.

By Behavior

Behaviors refer to quantifiable or observable actions, such as how often someone shops with or visits your farm, their level of engagement, and how much money they spend with your farm each year.

29. If you segment your customers into many different groups, watch out for people who fit into more than one segment. For example, you may have a segment for people who live within 10 miles of the farm and for people who shop with you at least one per week. If a customer falls into both categories, don't send them conflicting messages.

How to create a customer database

Most small and midsize farm use a variety of software programs and apps. Each program collects data in its own silo. It can be hard to integrate all your customer information into one database. Here are some ways to turn your existing tools into a more robust customer database.

Use your email service provider. Programs like Constant Contact, Mailchimp, and Klaviyo are small business communications platforms integrating ecommerce, social media, advertising, and SMS text functions.

Use a digital payments app or Point of Sale system like Square, Clover or ShopKeep. Many of these systems include email and texting capabilities, loyalty programs, and other services and tools that help you create a decent customer database.

Use a farm ecommerce program—such as Barn2Door or GrazeCart—or a general ecommerce program like Shopify.

Use a customer loyalty program. Both standalone loyalty apps and programs that are built into POS systems can track various types of customer data. But beware of additional costs they may charge each time you send a text or email.

Use a small business *Customer Relationship Management* (CRM) program, such as HubSpot, FreshSales, Zoho CRM, and Keap.

Frequency of purchase/visits

Regulars (visit at least once per week). Your regular customers are likely to be CSA members, weekly farmstand shoppers, restaurant or tasting room customers (if you serve food or alcohol), or workshop attendees. Regulars like special treatment; surprise them with a pound of fruit or a treat, solicit their feedback, or give them a heads-up when you've got a new product.

Occasional customers (visit at least once per quarter). Your occasionals may be busy, have limited incomes, or shop elsewhere. With the right messages and offers, you may be able to convert a few occasional customers into regulars; test some tactics.

Where to focus

MICROFARM MUSTS

You probably don't have a database. Think of each location where you sell as a customer segment. Compare sales patterns so you can customize your offerings for each segment. If more young children accompany their parents on Saturdays than during the week, offer kid-friendly snacks on the weekends.

SMALL FARM SMARTS

Make use of the data you've already got. If you have an email list or other customer database, sort customers by location, year joined, or other information you can easily access.

MIDSIZE FARM MAYBES

Analyze your data to identify all the customer segments you serve. Use your POS, loyalty program, or customer database to segment customers by products, services, or experiences purchased; purchase frequency, or amount purchased.

Annual customers. People who come for PYO or your annual festival may live too far to be regulars, yet may faithfully engage with you on social media. Some may not want to receive weekly or monthly emails; ask if they'd prefer an annual postcard or email reminder.

By level of engagement

You can also segment your customers by the frequency or level of their engagement with you. Someone may attend all your events, engage with you on social media, and post rave reviews of your farm on social media. Highly engaged supporters of your farm help increase your farm's visibility even if they're not regular customers. Message them to let them know you appreciate their support.

A small number of customers are responsible for the majority of sales for most businesses. This is commonly referred to as the 80/20 rule.

By total annual spend

Most businesses find that a small number of their customers are responsible for the majority of their sales. This is commonly referred to as the 80/20 rule: 20% of customers generate 80% of your sales. While the exact proportions may vary for your farm, you're likely to see this pattern.

Your biggest annual spenders may also be your regular customers or they may just spend a lot when they visit.

Top 20% spending customers. Your goal is to make these customers feel appreciated, not push them to spend more. Offer special perks that resonate with their interests and values. Give them a baseball hat imprinted with your logo. Ask them to be a taste-tester for your new yogurt flavor. Text to let them know about a new offering you think they'd like.

Top 21–40% spending customers. There are usually more people in this category than your top 20%. Some might spend more at your farm if you could meet more of their needs. Interview a few of them to ask what you could offer.

Increase sales

Make your farm and offerings look great

Simple changes to your signage, displays, packaging, or the layout of your farm or store can help you increase sales without raising prices.

Justamere Tree Farm is a sustainable maple farm in Worthington, MA that sells at local farmers markets and online. When I worked with them a few years ago, the owners were seeking to increase ecommerce revenue.

Although their ecommerce site was attractive, their product labels looked haphazard. Their products didn't look like they came from the same farm.

Justamere created a consistent design for their labels. As you can see from the before and after photos, their product line is more coherent. Sales increased after the redesign.

Any changes should make your farm and products more appealing to customers, create a more consistent image, or convey your brand in a more powerful way.

Here are three things to review when evaluating whether a design change might be valuable:

- Consistency: Are your logo, typefaces, and colors being used consistently? Do they look like they come from the same farm?
- Distinctiveness: Do your signage, products, and visual appearance stand out on the road, at your farm, at the farmers market, or on the shelves?
- Impact: Does your farm's brand personality and tone of voice reflect the feelings you want to inspire?

If you don't know, ask your customers.

Sustainable maple products from Justamere Tree Farm before and after the redesign. (Photo: Justamere Tree Farm)

Your target customers seek you out because they want to feel good about where their money goes.

Increase size of average customer spend per visit

Your target customers seek you out because they want to feel good about where their money goes. They want to help your farm stay in business and keep money circulating in your local economy. You may be able to increase their average spend per visit by encouraging them to shift where or how they buy their products, services, and experiences from an anonymous superstore to your farm.

Bundle and price your items differently

Matthew LeRoux, a marketing expert based at Cornell Extension, and his colleague, Todd Schmit at Cornell, conducted a 2018 study of customer spending and farmers' average daily gross sales at farmers markets.[30] The Ithaca-area vegetable farmers who participated in the study were required to use the Square app to record cash and credit card sales, so Matt and his team could analyze customer spending patterns.

The study found that the value of the average item purchased was only $3.38 and the average customer only purchased 1.8 items. The average volume may be higher in your area, but this seems shockingly low to me.

While the study's purpose was to uncover purchasing behavior, not test tactics, Matt and Todd shared ideas for how farmers could potentially increase minimum customer spend and/or average number of items purchased without raising prices. For example, they suggested that you pack two garlic heads together and charge $3, instead of selling a single head for $1.50/head,. They also suggested that you group similarly priced items next to each other, so that you can offer "Any 2 items for $7," "3 for $10." or even an "all this for $20" pack.

There are other ways that you can bundle your products differently. Many farmers package items such as tomatoes, berries, or peppers by the box or bag rather than by the pound. In addition to encouraging customers to buy more than one tomato at a time, it's easier

30. LeRoux, M., and Schmit, T. (2020, January). Electronic POS Bring New Opportunities. *Smart Marketing Newsletter.* https://dyson.cornell.edu/wp-content/uploads/sites/5/2020/01/Electronic-POS-Bring-New-Opportunities.pdf

for them to see how much a product will cost. It's also faster than having to weigh each item, saving time for you and your customers.

Display and suggest complementary products

Cross-merchandising refers to displaying or marketing complementary items together. For example, you might display all the products needed for making salsa on a table. Seeing the items next to each other can remind customers that they need both products.

Cross-selling—the classic McDonald's question, "Do you want fries with that?"—refers to suggesting that a customer add a complementary product. For example, you can ask if a customer wants a cup of coffee to go with that cider donut.

If your farm doesn't produce a complementary item that your customers are likely to want to pair with something you offer, consider partnering with another farm or business that produces a compatible product.

Provide convenience

The COVID–19 pandemic forced local farms and consumers to adapt their selling and buying behaviors to meet changing needs. Many local farms provided home delivery, online ordering, and curbside pickup for the first time. While it was painful to adopt new practices so quickly, both farms and customers were happy with the new options.

Improve your ecommerce sales

A growing number of farms operate ecommerce stores that ship meat, flowers, maple syrup or other items anywhere in the US. Ecommerce has its own set of tactics that will help you increase sales. Some farms won't offer free shipping to ecommerce customers because they believe it's not profitable. Research shows that 93% of shoppers will add items to their online cart or take other actions to quality for free shipping.[31] When calibrated properly, free shipping can increase profitability.

---

31. Utter, David A. (2018, January 13). Shippo Survey Results Highlight the Popularity of Free Shipping. *eCommerce Bytes*. https://www.ecommercebytes.com/2018/01/13/shippo-survey-results-highlight-popularity-free-shipping/

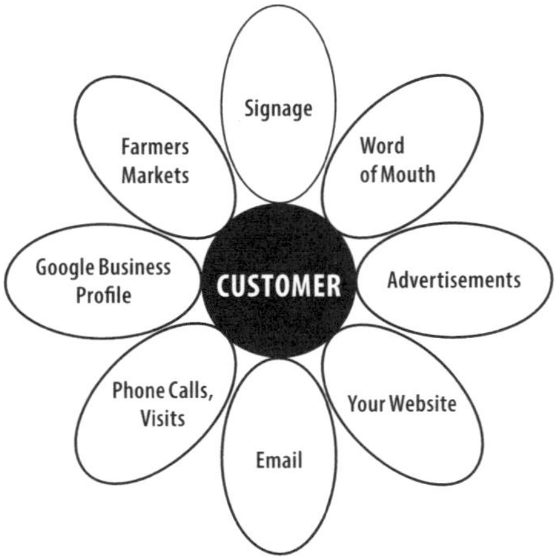

Customer touchpoints

Find new direct customers

Your farm needs a strategy for attracting new direct customers. No matter how much your customers love your farm, some will move away, change lifestyles, or get too busy to buy from you; you must find new customers each year. If you're trying to scale up your farm, figuring out how to acquire new customers is even more important.

It often takes multiple contacts—what marketers call *touchpoints*—with a potential customer before they make their first purchase, let alone become a loyal customer. If you're offering an expensive or complex service, customers want more information; people spend more time weighing options when choosing a wedding venue or CSA than choosing a restaurant.

The diagram to the left displays some common touchpoints. Prospective customers may encounter you online, in-person, or both, so your brand image and messaging should be consistent across all your touchpoints.

Where to focus

MICROFARM MUSTS
Adding a new sales location to an existing channel is usually the easiest way to add new customers.

SMALL FARM SMARTS
Test some new marketing tactics. If you're seeking more customers in a specific geographic area, try distributing flyers or running social media ads.

MIDSIZE FARM MAYBES
Sell more ready-to-eat foods, hold events, or add other enterprises. Promote them via direct mail or use online marketing tactics to attract new customers.

Attract new customers for each sales channel

Farmers markets

The advantage of selling at farmers markets is that it's someone else's job to attract a crowd each week. Once shoppers are there, your farm competes for attention with the other farmers and vendors. You need tactics that entice prospective customers to notice you, stop at your booth, and make a purchase.

Get noticed

Focus on your signage and displays. You need an attractive banner that instantly conveys what you offer. A colorful, well-organized display can help you catch someone's attention as they stroll by.

Be friendly. You and your employees should smile and make eye contact. If it's not too busy, step out in front of your booth and strike up a conversation.

Be busy. Crowded booths attract crowds. While everyone says they hate waiting in line, when they see customers lined up at your booth, they'll notice you and wonder what you're offering.

Get people to stop at your booth

Create an alluring presentation. Maximize the sensory appeal of products, so they look and smell enticing. Offer free tastes or samples. Keep restocking and culling so your displays are always full of high-quality items. Use bins, tablecloths, and props that reinforce your farm's brand, offerings and values.

Make it easy to shop. Even if it's crowded, people should be able to sense where to line up, walk through your booth, pick out product, and pay.

Stand out among the other vendors. Offer unique items or products that no one else at the market has. Introduce new cuts, flavors, or sizes. Ask people who stop by your booth if there are any hard-to-get products you could supply.

Make it convenient. If you have the capacity, sell prepared products to eat immediately or offer chopped, peeled, or heat-n-serve items

Get people to buy

Provide clear pricing and product information. Don't make us work to figure out what a product is or how much it costs. If your farm accepts SNAP or other benefits, post a sign. Let customers know what forms of payment you accept.

Stay friendly. Eye contact, smiles, and suggestions about how to cook, store, or use your products can help close the sale.

Offer pricing and buying options. Bundle, bag or box some items for a set price so they don't need to be weighed. Sell seconds or "ugly produce."

Extra credit

Market your presence at the market! Post photos on social media featuring what you'll be selling. Use the market's name, location and hashtags to promote your posts.

Community Supported Agriculture (CSAs)

Reaching people who already know your farm

People who currently buy any of your other products, services, and experiences are already familiar with your farm. When reaching out, always include an explanation of what you offer, how it works, and the benefits.

Pitch your CSA in-person. It's easier to sell something that may seem expensive or complicated, like a CSA, face-to-face.

Hand out CSA flyers or postcards at your farmstand, farmers market booth, PYO, and on-farm events.

Promote the CSA in your emails and social media. Ask your CSA customers to refer others and share your posts.

Hold a raffle for a free small share to help spread the word (and build your mailing list).

Reaching people who don't know your farm

Selling your CSA to non-customers is challenging. You have to attract their attention and, once you have it, persuade them to sign-up. As noted above, in-person contact is the easiest way to sell a CSA, because you can adapt your sales pitch to each person. Farmers markets, town fairs, and other local events are usually the best settings to reach customers, because people are there to explore.

Otherwise, look for a physical location where you're likely to encounter people in your target market, such as a local coffee shop or a yoga studio. Get permission to set up an eye-catching display of the wonderful foods in your CSA. Let curious people come to you; smile and make eye contact with everyone you can.

I don't recommend using paid advertising or direct mail to sell a CSA, since you'd need to spend a lot of money to reach enough people to make this worthwhile. If in-person contact isn't an option, partner with a like-minded business or organization willing to reach out to customers or members on your behalf. Many farms have successfully allied with health care organizations or large employers who want to make fresh local foods more accessible.

Attract new customers by location

Develop your strategy based on where your target customers are located. Do most of your customers live or work within 30–45 minutes of your farm? Do you attract day-trippers who come for PYO, events, or your award-winning hard cider? Or are you based in a tourist area that attracts one-time visitors from around the world?

How to attract customers using traditional tactics

Signage and flags. Physical or digital road signs on your farm, along roads leading to your farm, or on the highway are great ways for locals and tourists to find you. Flags and other things that wave or move, such as balloons and banners, are also good at attracting attention.

Table at local events, stores and coffee shops that attract your target customers.

Print material. Ask local businesses and nonprofits if you can leave flyers or postcards on the counter or post them on a bulletin board. Include a QR code to make it simpler for people to visit your website.

Partnerships and sponsorships. Associating your farm with like-minded businesses and nonprofits helps you reach like-minded customers.

Direct mail. Buying mailing lists, producing the materials, and mailing them can be expensive. If you have an eye-catching piece that's targeted to the right audience, however, it can be effective.

Print ads. There are fewer outlets for print ads these days. Some towns still have flourishing newspapers with loyal readerships. Print ads are most effective if you advertise regularly; you can also use them to promote a new enterprise or big event.

Radio ads and public radio sponsorships. Advertising on local commercial stations can be affordable. Public radio sponsorships can work well, but tend to be more costly.

Where to focus

MICROFARM MUSTS

Focus on attracting people who live near your farm. Locals are the most likely to become regular customers. Both traditional and digital tactics can be effective.

SMALL FARM SMARTS

People who live a few hours away from your farm can be good customers; traveling to your farm makes it more of a memorable experience. If they have a good time, they're likely to write rave reviews, follow you on social media, and recommend you to friends. Digital tactics are usually the most effective for reaching day-trippers.

MIDSIZE FARM MAYBES

Try to attract tourists for your events and agritourism activities. Use a mix of traditional and digital tactics to attract their attention while they're traveling in your area; follow up with digital tactics.

Find and keep wholesale customers

The best strategies for finding and keeping new grocery, food service, institutional, wholesale, and intermediary customers vary—as always—according to your farm's size, goals, and capacity, as well as the specific channels you want to pursue.

Most farmers refer to any of the above entities as a *wholesale* customer; therefore that's the umbrella term that I use throughout this section.

Here are the major channels to consider:

The common factor for all these channels is that you're selling or moving your goods through a third party.

Retailers

- Grocery outlets: Supermarkets, supercenters, mass market and club stores, natural food stores, convenience and drug stores, ecommerce sites, food coops, and independently owned grocery stores
- Food service outlets—chain and independent restaurants, sports arena, caterers

Intermediate markets

- Institutions—schools, colleges, universities, hospitals, prisons, and assisted living facilities, etc.
- Intermediary businesses—wholesalers, distributors, food hubs, and processors.

The common factor for all these channels is that you're selling or moving your goods through a third party. Someone else sells, prepares, or processes the food or products that consumers ultimately buy, eat or use.Here are a few things you should know if you want to sell to wholesale outlets.

- Identifying and reaching the person who has buying authority can be challenging
- You'll have to meet strict safety, handling, and quality standards
- You'll need to be on top of hidden expenses that come out of nowhere
- Other suppliers will be constantly vying to take your place

All wholesale outlets are not equal! Some wholesalers are independently and locally owned, while others are multinational corporations. Some will go out of their way to buy directly from farms, while others will only grudgingly give you a chance.

Start small and build your way up

If you're new to selling in wholesale channels, start with small, independently-owned restaurants, food coops, and natural or specialty food stores that seek to buy from farms. They're likely to be more flexible and helpful and tend to pay more than larger outlets. If your farm or product isn't a good fit for them, many provide some feedback as to why. Selling to just a few smaller outlets will help diversify your revenue sources.

If you're new to selling in wholesale channels, start with small, independently-owned restaurants, food coops, and natural or specialty food stores that seek to buy from farms.

Other sympathetic wholesale outlets include locally-based food hubs, local food aggregators, meal kit companies, and small distributors serving your state or region. Buying local is part of their business models. Once you're in, they'll often provide marketing and sales support.

Catching bigger fish

If you're already selling to small independents and are ready to scale up, go after bigger fish. Look for local institutions to contact, such as schools, colleges, and hospitals. Many institutions are committed to buying a certain percent of local food, so they're more likely to give you a try. Farm to School groups, Buy Local networks, food hubs, and other organizations in your area can help you reach out to institutions.

If you have sufficient volume and cash flow, approach natural food stores, supermarket chains, and club stores like Costco. Look for outlets with local food programs that offer more support and flexibility for small farms.

You can also approach grocery or food service distributors to carry your product. Distributors facilitate ordering, billing, storage and delivery for wholesale customers. Some wholesalers will only order from you if you're carried by one of their distributors.

Once a distributor agrees to work with you, you'll still have to go after accounts yourself (or hire a broker or outside sales rep to sell for you). Distributors offer opportunities for you to promote your products to potential customers, but you'll usually have to pay for any promotion.

Most of the tips that follow apply to any type of wholesale account. The bigger the channel, the more rigid they'll be about standards, terms, and requirements. Make sure you understand what you're getting into before signing any contracts.

Identify the buying process

Make a list of the wholesalers you want to target. Ask other farmers or local food businesses for suggestions. Visit locations to see what they're already selling or using in your category. See if they promote their local vendors.

Once you've identified the businesses to approach, find out how their buying process works.

Once you've identified the businesses to approach, find out how their buying process works. Processes vary, even for similar types of businesses. The person who makes buying decisions at independently-owned restaurants is often the chef; at some restaurants, the owner or a manager is in charge of purchasing.

What to find out

- Who decides which products they'll carry or suppliers they'll use? Is it a single person, such as the owner, chef, or department manager, or a high-level management team?
- What's the name and title of the person you need to contact? Are they the final decision-maker or do they just manage the decision-making process?
- What's the best way to contact them—email, text, mail, phone, dropping in?
- What's the best day of the week and time of day to reach them?
- How do they decide what products they'll carry or which suppliers they use? What criteria do they use?
- Do they consider new products and suppliers at any time or do they follow a schedule for reviewing each type of supplier or product? If they have an annual review for your category, when's the right time to contact them?

Some businesses post information about how to become a supplier on their website. If they don't, ask someone else who sells to that business if they'll share information. You can meet store buyers at an industry trade show, event, or meeting. Agriculture departments, business incubators, or Buy Local organizations hold events to help small farms connect with buyers. You can also call, drop by, or email to find out what you need to know.

Getting in

Wholesale buyers are motivated by their personal needs, desires, interests, and values, just like your direct customers. At the same time, they face pressure to maintain profitability for their department or business. Buyers are also concerned about price, quality, timely delivery, and whether your products please their customers.

Restaurants, supermarkets, and institutions operate on slim margins and face competitive pressure to keep costs down and prices low. They don't have a lot of staff, so buyers are juggling multiple tasks. Just like you, they're unlikely to be sitting at a computer reading email all day. Be pleasant and persistent.

Write a script or talking points before you contact a buyer. Practice before your first call. Compose something to say in case you need to leave a voicemail. Hopefully, your initial contact will lead to an in-person meeting with a buyer. Get your materials ready in advance. The chef/owner of an independent restaurant may only want a price list and samples; most businesses will want more written information, especially for packaged goods.

Materials to share with buyers

- Business cards
- Product samples
- Sell sheets[32]
- Promotional materials
- Farm photos

---

32. A *sell sheet* is an attractive, easy-to-read, one- or two-sided flyer, briefly describing your farm and products, emphasizing any characteristics that make them special. Include your logo, URL, contact information, product photos, prices, and UPC codes.

Be prepared to talk about how long you've been farming, your growing/production and food safety practices, current customers, product availability, packaging, distribution and delivery area, as well as your margins and suggested retail pricing.

Study your buyer's website, social media, and available information in advance. If the buyer is ready to move ahead, be prepared to ask your own questions.

Questions to ask

- What are your requirements for my products (sizes, packaging, ripeness, cuts)?
- Will you buy direct from me or do I have to work through a distributor?
- What's the best delivery schedule?
- What are your terms?
- What are your insurance, food safety, and packaging requirements?
- How do you place orders?
- Which specific products are you potentially interested in? Do you have any idea of the volume you're likely to order?
- What's the next step? How and when should I follow up?

When dealing with larger businesses, it can take months to schedule a meeting, be approved, and get your products into the store. The business may also impose more requirements on you—free shelf fills, mandatory promotions, buy backs, and other expenses. Review every requirement with the help of an accountant, lawyer, or industry professional.

If you get turned down, try to find out why. A no is often just a "not now." Ask if you can recontact them in six months or a year.

If you get turned down, try to find out why. A no is often just a "not now."

Staying in

Getting into a store or restaurant is often easier than staying there. If your products don't maintain their quality standards, aren't delivered on time, or simply don't sell well, there's always another supplier waiting to replace you.

In most cases, the buyer who got you in will also be responsible for ordering; if not, find out the name and title of your contact person. Some businesses use an online ordering system with no personal contact; others will want to email or call in their order. Some need weekly sales calls to prompt them to check their inventory. Ask how they want to be contacted and the best times to reach them.

Get to know the buyer at each account. If you build a relationship with them, the interaction will be more satisfying for both of you. They'll picture you and your farm when placing or receiving an order. Share photos and videos. Invite the buyer and their colleagues for a farm tour. Try to meet as many employees as possible, so they can share their excitement about your farm with their shoppers or eaters.

Get to know the buyer at each account. If you build a relationship with them, the interaction will be more satisfying for both of you.

Don't compromise your principles to appease a big customer. You offer something of great value. Your products have tremendous appeal for their customers and help differentiate their business from their competitors.

Out of sight, out of mind

Stay on top of how your products are being handled, stored, and displayed everywhere you sell. Unless you make direct deliveries to your customers, you won't know how they're handling your products. The school buying your root vegetables may have inadequate storage facilities. The small store may not be culling produce frequently enough. You or someone representing your farm should visit your wholesale customers to check things out at least quarterly or biannually.

The businesses you sell to are focused on pleasing *their* customers, not their suppliers. The businesses decide if, when, and how your farm will be visible to their shoppers and eaters. If promoting you as a local farm makes their customers happy, your wholesale customers *may* be willing to give you visibility, but you can't count on this.

If a supermarket sells your sweet corn, their signage is likely to say, "Local sweet corn," not "Sweet Corn from Farmer John, Peabody, MA." This streamlines their operations. If they have a custom sign for your corn but run out of your product, they'll use another supplier's corn; someone will have to remember to switch the sign. If they stick with a generic sweet corn sign, they're covered.

If you supply a packaged product, you'll create the packaging (unless you're providing a private label product). Your goods will be branded with your farm's name. The store controls the use of point-of-sale materials, shelf-talkers, and promotional opportunities. Unless you're able to negotiate good shelf placement or pay for promotions, shoppers may have trouble finding you.

Businesses that pride themselves on buying local will go out of their way to let their customers know who produces their food.

Businesses that pride themselves on buying local will go out of their way to let their customers know who produces their food. Food coops and specialty food stores will proudly spotlight local vendors. Many farm-to-table restaurants print the names of the farms they buy from on their menu or blackboard. Some institutions use signs on cafeteria tables, employee newsletters, and social media to promote their local vendors. Choose wholesale customers who will be an ally for your success.

Reaching the shoppers and eaters

Increase your visibility. You can't count on wholesale accounts to promote you to their shoppers and eaters, Look for ways to make customers at each store, restaurant, or institution aware of your farm. This visibility will drive demand for your products. It also increases your overall brand awareness, helping you to build sales and customer loyalty in other channels.

On-site tactics

Signage and point-of-sales displays. See what type of materials other suppliers are using. Ask your buyer what types of signage or displays they permit. Some accounts will let you supply your own signage and display materials (within their guidelines), while others will make signs for you (usually for a price).

Packing materials and QR codes. If you create attractive packing materials, such as printed boxes or bags, some businesses will incorporate these into their displays.

Red Tomato in Providence, RI, worked with growers and a key customer to create a custom box for shipping and displaying heirloom apples utilizing sustainably sourced cardboard and ink.
(Photo: Red Tomato)

- Supply stores with apple bags printed with your farm name and logo.
- Kitchen Garden Farm in Sunderland, MA prints their farm name on the twist ties they use to bind their produce. The twist ties also identify their products as organic.
- Red's Best, a seafood distributor and ecommerce business that buys directly from local fishermen, packs every piece of fish with a QR code that details who caught the fish and where it is from. This transparency adds value to the product for the retailer, while helping to build Red's Best's brand.

Product sampling, demos, tastings, and tabling. Take advantage of opportunities for shoppers or eaters to meet you, taste your products, and learn about your farm. Some venues encourage farms to do their own demos, while others use their own staff and charge you a fee. Be sure to budget time and money for supplying product samples and labor. It usually takes more than one taste or encounter to build a connection with a shopper or eater, so plan to return to your key outlets multiple times.

Promos, displays, and merchandising. Most stores, restaurants, and institutions offer seasonal events, displays, and specials featuring products, recipes, and gifting opportunities designed to meet their customers' needs—and put them in a good mood. Be creative about suggesting ways to use or feature your product. Businesses work on annual promotional calendars with long lead times. Talk to your buyer about the best timing to present your ideas.

Many store chains and distributors require accounts to offer trade promotions a certain number of times per year. Putting your products on sale does attract attention and encourage people to try them. This may not turn them into regular customers; it can train shoppers to wait until your product is on sale to buy again.

Plus, some stores and distributors take advantage of trade promotions to stock up on your product at the reduced price. When the sale period is over, they go back to selling any excess at the regular price and pocket the difference. Be on the alert for significant overbuys and talk to your customers about minimizing this practice.

Off-site tactics

Promoting your presence at any stores, restaurants, and institutions through your own communications channels is one of the most effective tactics at your disposal. You control these messages (unless your contract with your customer restricts what you can say). You need to have an active online presence to be successful. In addition to helping increase your product sales at those outlets and keeping on their shelves, it demonstrates your marketing muscle.

Website. Create a *locator* page on your listing all the places your products are available. Include a page or link for wholesale inquiries.

Social media. Post product availability, demos, sales, and any information that will appeal to your social media followers.

Email. Many of your direct customers at farmers markets or in your CSA also eat at local restaurants or shop at other stores, especially off-season. Share stories and links in your email about where they can buy or eat your products or see you at events.

Outreach to traditional media, bloggers, and influencers. This includes articles, calendar listings, social media posts, and mentions of your farm in print, on radio, TV, videos, and podcasts. Some businesses want to retain tight control over how their brands are depicted, so ask your wholesale customers if you need prior approval to mention them.

Reach out through your other channels. Promote your presence at stores, restaurants and institutions through your farmstand, farmers markets, PYO, and other events.

Grow your website

Your website is your digital home

Websites used to be the primary way to market your farm online. Creating a website was costly and cumbersome. You had to know some html coding and have graphic design skills to build a site. Not surprisingly, relatively few farms had websites.

Today, there are hundreds of website platforms, templates, and tools that enable you to build your own site or to hire someone to create a basic one-page site for a few hundred dollars.[33]

While you can promote your farm through many other online channels, such as social media, directories, and review sites, websites are still an important way to market your farm. Social media can help you connect with customers, but those channels are in business to make money, not serve your farm.

Your farm needs a website that serves as your digital homebase where you store articles, images, videos, calendar listings, and other content. When you have a paid website and domain name, you own and control your URL, the content, and branding.

Websites and social media complement each other. If your customers spend time on specific social media channels, your farm should have a presence there. When you post on social media, link back to your website when you can. This drives traffic to your site, which helps you get higher search engine rankings; it's a virtuous circle.

For suggestions on how micro-, small, and midsize farms can make effective use of social media, see the *Grow your social media* section.

Start with a one-page website

Don't have the time, money, or expertise to create a full website? Start with a one-page site. Use Google, MailChimp, Wix, or another platform to set up a free or inexpensive website. They offer templates, so you don't need to know any coding or graphic design.

Where to focus

MICROFARM MUSTS

If you have a Google Business Profile or a passive social media presence, you can get by without a website for a while. Create a basic one-page site when you can.

SMALL FARM SMARTS

You need a four-page or full-size website.

MIDSIZE FARM MAYBES

You need a full-size, visually-compelling website to help differentiate your farm. Showcase your purpose, offerings, and anything that makes you makes you special.

33. Expect to pay $300–$500 to hire someone to create a one-page website, $500–$1500 for a four-page site, $1500–$5,000 for a full site, and up to $10,000 or more if your site is more complex.

DIY vs. hiring a consultant

Should you create a website yourself or hire someone to do it for you? If you're comfortable using social media or email programs like Constant Contact or Mailchimp, you can easily create a one-page website using one of the free templates.

Need a four-page or larger website? If you're reasonably tech-savvy and have text and photos ready to go, you can build an attractive four-page or larger website using any of the website-building platforms.

If you'd rather spend your time farming than creating a website, hire a web developer, graphic designer, marketing consultant or agency. Individual consultants with the right skills are hard to find, but tend to be less expensive than agencies.

Simply enter your headlines, text, images, and other elements into a template, Tell people who you are, what you offer, and where you're located. Say a bit about your purpose, values, and farming methods. Include your farm name, logo, descriptor, and other brand elements. Provide your address, hours, and contact information. Hit "Publish" and you've got a website.

Free sites are great when you're starting out, but they do have limitations. You may not be able to use your own domain name. Customer support may be limited. The website platform could change its terms and start charging you for the site. When you're ready, upgrade to a paid site with your own URL.

Expand to a four-page website

As your farm grows, you'll need more than a one-page site to attract new customers and keep current customers informed. You can create a low-cost, effective website featuring the following four pages:

Home page: A quick, compelling overview of who you are, where you're located (or where your products are available), and what you offer. Your home page should also express your values or purpose.

Offerings: Describe your products, services, and experiences.

About: One or more pages that tell your story, including who is farming, how and why you farm, and any relevant information about your location.

Contact: Include your contact information, a Google map with directions to your farm, and links to any social media channels where you have a presence.

Include your farm name, address, phone number, email, and social media links in a header or footer that appears on all pages. This makes it easy for visitors to find this information and can help you show up higher in search engine rankings.

Create a full website

If you're a midsize farm, have multiple enterprises, operate an online store, or are actively seeking to attract new customers, you'll need a full website. If you haven't touched your site in the past two years, it probably needs an update.

A full website is an empty canvas. It can help attract new customers, increase sales, build customer loyalty, differentiate your farm, and advance your mission. What you do with it depends on your farm's goals, target customers, and available resources.

An effective website should be easy to navigate, offer compelling content, be mobile-responsive, be secure, and attract and persuade target customers to connect with your farm. We'll discuss these elements in the rest of this section; the last topic is covered in *Get found online*.

An effective website should be easy to navigate, offer compelling content, be mobile-responsive, be secure, and attract and persuade target customers to connect with your farm.

Be easy to navigate

Your website layout should be simple and clear. Visitors should be able to find what they want in a few seconds using no more than two or three clicks. If you make them wait too long or work too hard to get to the content they want, they may give up and go to another site.

Each page of your website should ideally contain 300 words of more about a unique topic. If you have sufficient content on related topics, create subpages for each subject. For example, if your orchard offers PYO apples and peaches, you could create separate subpages for PYO apples and PYO peaches. The main PYO page could provide overall information about your orchard; the PYO apples and PYO peaches subpages could describe your varieties, state when the fruit typically ripens, and post recipes.

Having a separate page for each popular topic makes easier for people to find your site when they do an online search for that subject. It's also helpful for search engine optimization. (See *Get found online* for details.)

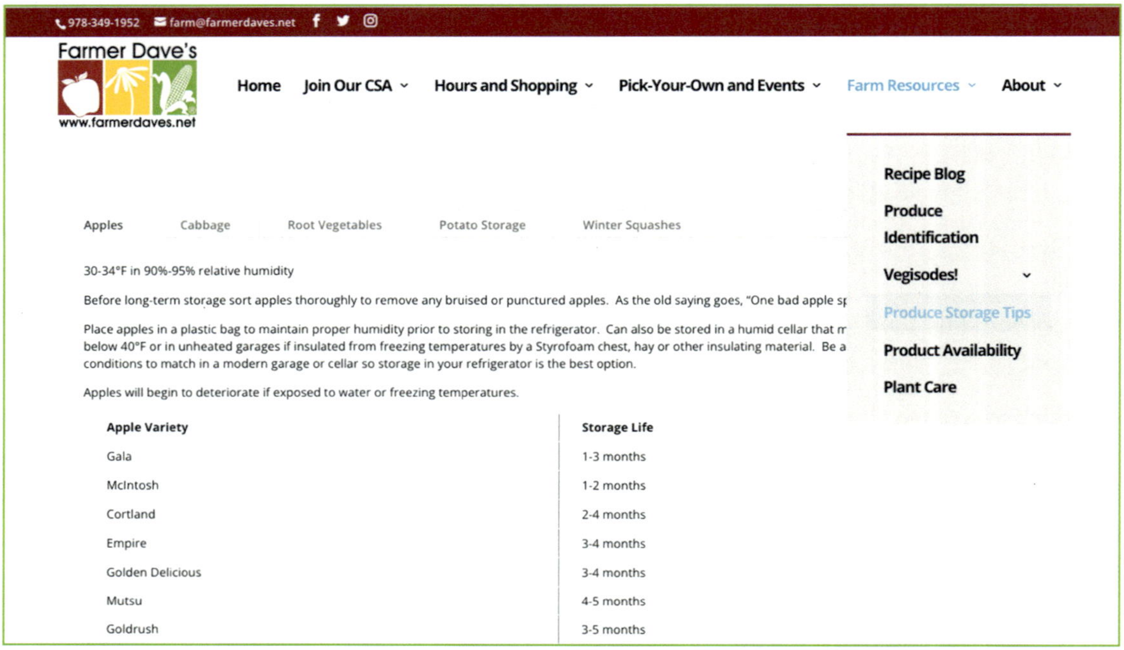

Screenshot of how Farmer Dave's uses a secondary menu on his website.

The main navigation menu on your website should feature the most important sections of your website. The main menu should have no more than 3–6 items, so that it's easy to view at a glance. Smaller navigation menus also work better on mobile devices.

If your site includes subpages under each main menu item, create a dropdown menu so visitors can click or "mouse over" each item to see the additional topics.

You can add an additional—*or secondary*—navigation menu to any page. For example, Farmer Dave's in Dracut, MA uses a secondary menu on one of the sections of his website. The top navigation menu includes a useful catch-all label titled "Farm Resources." The Farm Resources menu leads visitors to six different subpages, including one titled "Produce Storage Tips." The Produce Storage Tips subpage contains a secondary navigation menu with tabs. Users can click each tab to view storage tips for the produce category of their choice.

Include internal links (or clickable buttons) to other relevant content on your site. These links can lead visitors from your main page about a topic to relevant subtopics, or take them back to the main topic from a subpage. For example, if you're describing the different type of CSA shares you offer, link to your CSA registration page and link back to your main CSA page from the CSA registration page. These internal links can help your top-level pages rank higher in search results.

Offer compelling content

If you interact with customers regularly, you probably have a good sense of types of stories, images, and information that your customers value.

- What questions and comments do you get when you interact with your customers?
- Which pages on your website get the most visits?
- Which social media content gets the most views, Likes, comments, and shares?

Your text, headlines, and images should appeal to your current customers *and* help you attract new ones.

Be mobile-responsive

Your website must look good and be easy to use on a mobile phone or tablet. This means not just being *mobile-friendly*, but being *mobile-responsive*. The terms are often used interchangeably, so it can be confusing.

Mobile-friendly sites can be used on mobile devices, but may not be designed with mobile in mind. Since mobile screens are smaller and differently shaped than desktop or laptop computers, mobile-friendly designs are typically shrunk to fit whatever device the viewer is using. It can be frustrating to use a mobile-friendly site, because the text tends to be small and the buttons and links don't have enough white space around them to make them easy to click.

Mobile-responsive sites have layouts that automatically adjust to the size and shape of the user's device. Everything is easy to use and looks good on all devices.

When possible, start your website design by envisioning how it will work on a mobile device, then consider how it will function on a desktop computer. Most website platforms offer mobile-responsive templates that make it easy to create a mobile-responsive site.

Ask a few customers to test your design on a mobile device to see how it works. You can run a free test to rate the mobile-friendliness of your site.[34]

Load quickly

Each page on your site should ideally load in 2–3 seconds. Most small websites should have no problem meeting this goal, but if your site has a lot of large images, it can slow down the load time. You can use software to minimize the size of your images so they load more quickly.[35]

Be secure (encrypted)

Most websites that have been built or updated in the past few years are protected with an SSL certificate, which means that any data passed between web servers and clients is encrypted. Secure sites start with https, not http. Credit card data or other personal information is more likely to get stolen from sites that aren't encrypted.

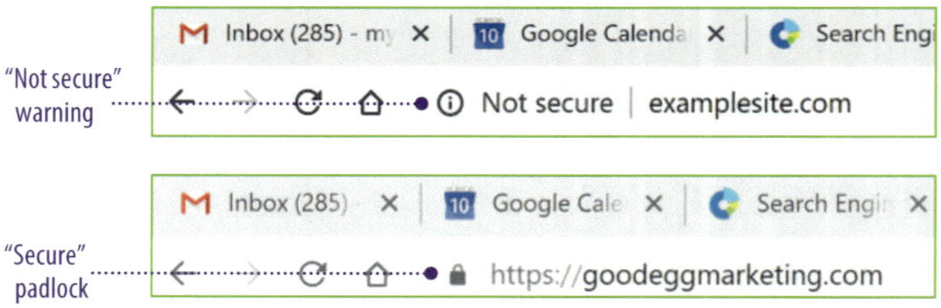

Sites protected with SSL technology have a little padlock next to their URL

The Google Chrome browser displays a little warning in front of any site that's not secure.

In addition to protecting your web visitors, Google tends to give secure sites higher search rankings. If you have an older site that isn't secure, a web developer should be able to migrate your site to a secure setting for less than $200.

---

34. Do a web search for "Google mobile-friendly test tool " to get a mobile-friendliness rating and tips for how to improve it.

35. Do a web search for "test website load speed" to find a free site where you can test your site speed and get tips for how to decrease your load time.

Grow your social media

Social media *can* be a great way to attract new customers and build customer loyalty. It can *also* be a huge time sink without much payoff.

It's not you, it's social media itself. The platforms seek to attract as many eyeballs as possible, so they can sell ads that reach those eyeballs. Unless you have the time and skill to create posts that inspire a "Wow," "Cool," or "Aw . . .," it can be hard to get more than a few Likes without investing in social media advertising.

At the very least, having a social media presence will help your farm get found online. At best, it can attract people to your farm for events, PYO, and your farmstand; stimulate sales; and build your brand. As with any tactic, you must adapt it to fit your farm's goals and resources.

Where to focus

MICROFARM MUSTS

Start with a passive presence on at least one social media platform. Become active when you're ready.

SMALL FARM SMARTS

Maintain an active presence on at least one social media channel.

MIDSIZE FARM MAYBES

Set up a Facebook group, pay to boost posts, or run social media ads to get more out of your social media presence.

Social media for farmers who hate social media

Having a social media profile will help your farm get found online. Even if you hate social media, it's worth having a passive presence on at least one channel, especially if you don't have a website or Google Business Profile. When someone enters the name of your farm into a search engine, your social media profile is likely to appear in the top results. Voila! Instant online presence.

A "passive" profile is one that you don't actively use. Create a complete profile and *pin* a friendly, warm post to the top of your feed.[35] It can say something like: "Hello! We're busy farming [whatever you offer]. We set up this profile to help you find us, but we don't monitor this page. To learn more about our farm, please visit [in-person, at a farmers market, or wherever you want to see them] or contact us at [provide an email address or phone number or address]."

36. You can *pin* a specific post so it always appears at the top of your feed. Do an online search to find instructions for how to pin a post on a specific social media channel.

How to choose a social media channel

Each social media channel tries to establish unique, useful, and entertaining characteristics that will appeal to specific types of people. Unless social media is your favorite hobby, choose one or two platforms where you feel comfortable and that are likely to attract your target customers.

Follow other local farms to see which channels they're using, what they're posting, and what type of responses they're getting. Ask other farmers which social media channels are most effective for them.

Do an online search for "choose the right social media platform" for up-to-date information on each channel.

"Lurk" before you work. If you're not familiar with a channel, sign-up for a personal account and watch what type of content is getting shared before you make a single post. It may be tempting to jump on the trendiest channel—right now it's TikTok—but if you don't have the time and skill to create funny short videos, don't bother.

Choose the platforms that your current and prospective customers are most likely to use. For most farms, that's still going to be Facebook and Instagram. If you don't know which social media channels your customers are on, ask them.

If you're social media averse, stick with the app you're most likely to use.

Get more mileage from your Facebook page

Facebook is the largest social media channel, with over 267 million monthly active users in the US and Canada in 2021.[37] Originally created for college students, Facebook is commonly seen as Baby Boomer territory now. While Facebook's teenage users have been plummeting, 87% of millennials still use Facebook at least once a week.[38]

Outwit the Facebook algorithm. Facebook wants people with business pages to pay for advertising, so it keeps tweaking its algo-

37. Iqbal, M. (2022, June 30). *Facebook Revenue and Usage Statistics (2022)*. Business of Apps. https://www.businessofapps.com/data/facebook-statistics/

38. Digital Media Ninja. (2021, May 25). *What social media do millennials use?* https://digitalmedianinja.com/what-social-media-do-millennials-use/

rithm to show your regular (unpaid) posts to fewer and fewer people. As of this writing, only 3–6% of your posts are likely to get displayed in your fans' news feeds. That means that if you have 100 fans, only three to six of them will see a typical post. Don't despair. Develop compelling posts that inspire your followers to comment and share; Facebook does display posts with lots of engagement to a higher percentage of your fans.

Show, don't sell. Most people follow farms on social media because it evokes warm feelings of being at your farm, interacting with you and your staff, seeing how their food is grown or produced, enjoying the views, and feeling the wind on their faces. While most of us non-farmers romanticize farm life, don't be afraid to talk about the challenges. Share stories, videos, and photos that help your fans understand life on your farm.

While most of us non-farmers romanticize farm life, don't be afraid to talk about the challenges.

Try out different types of content to see which ones get the best responses and do more of those. Animal videos are always a hit!

Follow other pages and share their content. Follow the Facebook pages of other farms, food businesses, food nonprofits, environmental groups, and relevant media outlets. Schedule 10 minutes a day to go through your Facebook feed. Comment on and share posts; when you engage with other peoples' content, they're likely to reciprocate.

Use Facebook Stories, especially for behind-the-scenes photos and videos, especially funny stuff. Stories are also great for promoting daily specials or events. Even though Stories only show up for 24 hours, they get featured at the top of your fan's newsfeed, so they're more likely to get seen. Create Stories using the Facebook mobile app, which provides you with some fun features like stickers and polls to gussy up your posts.

Use Facebook Events to promote events. Group a series of programs or classes together into one posting. Anyone who clicks Interested or Going for one of these programs will get notifications about each event in the series.

Try Facebook Live. If you have a cell phone, you can stream from anywhere. No one's expecting Hollywood quality. Do plan what you'll show and talk about. Announce your Facebook Lives in advance, so your fans can plan to tune in. Download your Live videos so you get views afterwards.

Create a Facebook Group. Many farms have gotten better results from their private Facebook groups than their public Facebook posts. Group members get notices when there's a new post. And people are more likely to post, share, and comment in a private group than on your public page. Groups work especially well for CSAs or small communities where people know each other. But you can also offer a Facebook Group around a specific customer interest, such as gardening or cooking.

Don't feel compelled to post every day, unless you're promoting daily specials or PYO conditions.

Focus on the quality of your posts, not the frequency. Don't feel compelled to post every day, unless you're promoting daily specials or PYO conditions. Do try to follow a predictable schedule, so your fans know when to expect your posts. Posting 3–4 times a week or more is ideal for most farms.

Respond to all comments. Schedule a few times a day to go into your Facebook page and respond to comments. If appropriate, ask the commenter for more detail or ask if anyone else feels the same way, so you can get a dialog going.

Take the stress out of developing new content ideas. Schedule particular types of posts for specific days of the week. For example, you could post a recipe on Sundays or Mondays, a story or video of what's going on at your farm on Tuesdays or Wednesdays, and a preview of what's in season or available on Fridays.

Save time by creating a week or month's content in advance. Use Facebook Business Suite or third-party apps like Hootsuite or Buffer to schedule your posts to appear on your designated days.

Include a Call to Action on your page, such as an email sign-up form or link to a video about your farm.

Upload videos directly to Facebook. Facebook wants your fans to spend as much time as possible on Facebook, so its algorithm prefers *native* videos. If you have a video to add to your feed or stories, upload it directly to Facebook, rather than posting a link to YouTube or another channel.

Get more mileage from your Instagram page

Instagram is a social media channel designed for sharing photos, videos, quotations, and other visuals. Although it can be used on a desktop computer, it's primarily a mobile app.

Instagram can be an effective way to attract new followers and appeal to customers under 45. Instagram's overall organic reach is said to be just 3–6% (similar to Facebook), yet many farms are exceeding that percentage.

Like Facebook, Instagram can be a time sink. If your main goal is to stay in touch with your customers, you don't have to post every day in order to have decent visibility with your current supporters. Do try to post at the same days and times every week, so your followers know when to expect your posts.

If you enjoy taking photos and videos on a cell phone, having an active Instagram feed can be managed on 15 minutes a day. Here's the workflow:

Take photos throughout the day as something catches your eye, then review your photos at a regular time (either daily or a few times per week).

When you want to post, pick out an appealing image—crop, filter or edit if necessary—then upload it directly to Instagram or to an Instagram app like later.com.

Add a caption and hashtags (for your location, women farmers, young farmers, the topic of your photo, or whatever's relevant) and post.

As with any social media channel, you'll get a lot more from Instagram if you plan your posts. Think about which content is likely to appeal to your target customers and brainstorm how to provide it.

While people who follow you on Instagram will mostly see individual posts, Stories, or videos, when someone visits your Instagram page, they'll see your nine most recent posts laid out in grid form in descending chronological order. Therefore, think about how your posts will look when viewed in your grid. Use a consistent visual theme that reflects your farm's brand personality.

Good Egg Marketing Instagram grid.

The image above shows a sample grid from the Good Egg Marketing (@goodeggmarketing) Instagram page. While not perfectly balanced, there are some common elements that reflect Good Egg's brand style and values:

- I use a white background for most images so the bright colors will pop
- I display photos that reflect the cultural and ethnic diversity of farmers, food entrepreneurs, and foodies in the industry
- I intersperse quotes among the images

Instagram tips

Write engaging captions. People have to click on your image to view the caption, so why bother writing them? Educational, funny, or inspiring captions encourage people to react to and comment on your posts. Longer captions typically attract more engagement. While you can write up to 2200 characters in a caption, the length of your caption should match the content and tone of your post.[39]

Use Instagram Stories. Stories are special types of posts that only live for 24 hours. Instagram users prefer scrolling through their feeds to watching Stories. So why bother creating Stories? They're a great way to provide a quick glimpse of what's going on at your farm. Stories don't have to be polished; users expect them to be rawer than your posts. Stories have lots of cool interactive features like stickers that can be used to engage your followers. Archive your Stories under Highlights so they're permanently available.

Use hashtags strategically. An Instagram hashtag is a clickable keyword or phrase proceeded by the hash (#) symbol. Include hashtags in your posts or stories to help people find and follow you on Instagram. For example, if your post or story includes the #cider-slushy hashtag, everyone can see your post when they scroll through the #ciderslushy hashtag feed: some may decide to follow you. Location hashtags—such #easthamptonma—are another way to help people find you.

You can post hashtags in your caption or in a comment. Choose hashtags that your target customers are most likely to follow; check out other local farms to see which hashtags they're using. Vary some hashtags in each post so you keep reaching new people. Most people just use just a few hashtags You can use up to 30 hashtags in a post, up to 10 in a story. Most people just use a few hashtags, but 11 hashtags per post is considered optimal. Test to see which hashtags and how many work best for your farm.[40]

---

39. Chacon, B. (2021, November 24). "What is the best Instagram Caption Length?" LaterBlog. https://later.com/blog/instagram-caption-length/

40. Newbury, C. (2021, July 19). "Instagram Hashtags 2022: The Ultimate Guide." Hootsuite Blog. https://blog.hootsuite.com/instagram-hashtags/

Use multi-image posts. Multi-image posts can include up to ten images or videos (with one caption) for fans to swipe through. Multi-images (aka carousels) get more engagement than single image or video posts.[41]

Use an app that lets you share more than one link in your bio. Instagram doesn't permit you to post links in captions. You can share multiple links in your bio using free/low-cost apps, such as Linktree or Later. This enables you to post direct links to events, your CSA, or anything you want to highlight.

Use videos. Instagram prioritizes video posts over still images. The app is constantly changing the rules for where and how you can post videos; search online for the latest options. Viewers don't expect your farm to produce high-quality videos, as long as your content is educational, fun, or inspiring. Add filters, captions, interactive backgrounds, stickers and other elements to your videos.

Follow other accounts, including anyone who follows you. Many of these accounts will like and comment in return.

Share the love. Open Instagram at least a couple of times a day. Scroll through Stories and your feed, liking and commenting on other Instagram posts. Follow other accounts, including anyone who follows you. Many of these accounts will like and comment in return.

Share your Instagram feed. Facebook lets you link your Instagram and Facebook accounts, so that all your Instagram posts automatically post to Facebook. If you do this, post your hashtags in comments, rather than in your captions, so they don't clutter up your Facebook post. Some third-party social media apps let you customize your posts for each channel before sharing them, which I find more efficient.

Embed your Instagram feed on your website. You can embed your Instagram feed on your website so that it displays your latest posts. This is a great way to show off the content you worked so hard to create. It also encourages more people to follow and engage with you on Instagram.

---

41. Southern, M. (2020, August 30). "Instagram Carousels Are the Most Engaging Post Type [Study]." Search Engine Journal. https://www.searchenginejournal.com/instagram-carousels/

Get found online

Every farm should have an online presence. Current and potential customers must be able to find you when doing a web search, looking for your address, or asking Siri to "find an apple orchard near me."

Any size farm can create an online presence using a Google Business Profile, social media presence, or one-page website.

Showing up on the first page of Google when someone is searching for a product, service, or experience you offer is usually more difficult. Lots of other farms, businesses, and organizations are competing for just a few slots.

You can use many tools and techniques to get found online, such as social media, getting listed in online directories, search engine optimization, and online advertising. This section walks you through these tactics.

Online Marketing Tactics

FOR ALL FARMS

- Google Business Profile
- Social media presence (passive for microfarms, active for small farms, boosted for midsize farms)
- Website (one-page for microfarms, four-page for small farms, full site for midsize farms)

ADD > FOR SMALL + MIDSIZE FARMS

- Get listed in online search directories or on review sites
- Analyze your website metrics (see *Measurement*)
- Adopt basic search engine optimization tactics

ADD > FOR MIDSIZE FARMS

- Online advertising

Where to focus

MICROFARM MUSTS

Develop an online presence. Use simple, free/low-cost tools, such as a Google Business Profile, a passive social media profile, or a one-page website.

SMALL FARM SMARTS

Improve your online visibility. Keep your Google Business Profile updated, optimize your website to get found in local search results, and have an active social media presence.

MIDSIZE FARM MAYBES

Attract new customers with digital marketing. Optimize your website to get found in organic search results, boost some social media posts, and consider other online advertising.

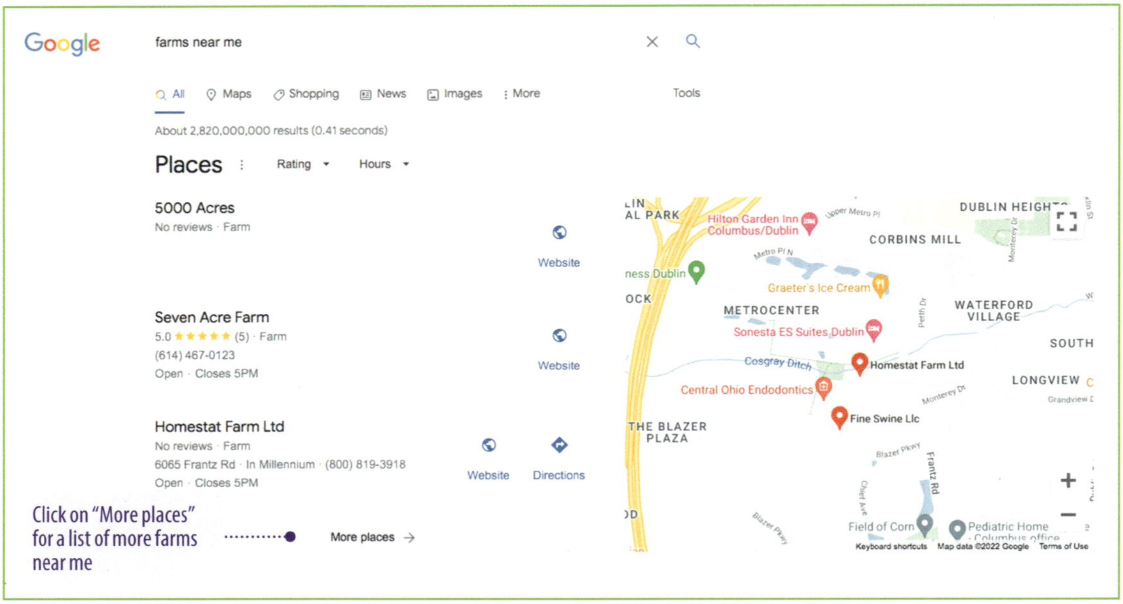

A screenshot of a search for "farms near me" shows how Google displays the Local 3-Pack and map.

Get found in Google's local search results

When someone enters a query into the Google search box, Google displays different types of results. Depending on your search, Google is likely to display paid ads, various information boxes (such as *rich snippets*, definitions, and answer boxes), and regular (unpaid) website listings.[42] Your website is potentially competing with every other website in the world to get one of the coveted free listings on the first page of Google search results.

If Google detects that someone is searching for a nearby business or service, it displays local businesses *before* all the regular listings from other websites. The three most relevant businesses appear in a *Local 3-Pack* (see the above screenshot). Google also displays a *Map* with a pin for the top locations. Getting featured in the Local 3-Pack is a tremendous advantage for your farm!

Google reportedly weighs 200+ factors when determining which web pages are the best match for each query. How does Google decide which listings to feature in the Local 3-Pack?

How search engines work

Search engines like Google, Bing, and DuckDuckGo continuously scan websites, directories, images, databases, and other online data, indexing the information. Each search engine uses an algorithm to analyze your search query, match your request to the indexed information, and determine which results are most likely to satisfy your intent.

Since 87% of US online searches take place on Google, I'm focusing on how to get found there, but most of these tips apply to any search engine.

42. Google refers to non-local, unpaid listings as *organic search results*. To avoid confusion, I'm using the term *regular search results*.

How to create a Google Business Profile

Enter your farm name into the Google search box. Click "Claim this business/ Manage now." Verify that you're the business owner. Google will email or physically mail you a postcard with a verification code.

Fill out your profile as completely as you can. According to Google, "Customers are 70% more likely to visit and 50% more likely to consider purchasing from businesses with a complete Business Profile."[43]

Use the "Q&A" feature to pose and answer frequently-asked questions.

Add photos and videos.

Choose your business categories thoughtfully. If your primary category is "farm store," and secondary category is "orchard," you'll be more likely to appear in searches for groceries than PYO; which category is more important for your business? You can choose up to nine secondary categories; focus on your top four.

Pay attention to reviews. Getting many 4- or 5-star reviews can help your search rankings. Encourage customers to mention a favorite product or service in their reviews—"Love their fresh non-GMO corn!" or "My kids look forward to the hayrides every year."—as those mentions can help you appear in searches for those offerings. Respond to all reviews (whether positive or negative), because Google also weighs this factor when ranking sites.[44]

The Google algorithm is top-secret, but local search experts say that the contents of your Google Business Profile (GBP) have the most impact on whether you make it into the top.[45] Google Business Profile—formerly known as Google My Business—is a free online listing where you can post information about your farm's offerings, location, hours, contact info, photos, and events.

---

43. Complete your Google Business Profile on Google (2022). Google support. https://support.google.com/business/answer/10515606?hl=en

44. Responding to all reviews will please your customers in addition to helping with Google search results. Research shows that people are more likely to review businesses that comment on all reviews. Besides, it's an opportunity for you to thanks people for positive reviews and address any issues that arise. If you believe that someone submitted a fake review, search for "how to remove reviews from your business profile on Google" online to learn how to contest it.

45. Coe, A. (2021). *The Top Local SEO Ranking Factors in 2021*. Bright Local. https://www.brightlocal.com/learn/local-seo/how-to-improve-local-seo/ranking-factors/

Other factors that can help you get into the Local 3-Pack include your farm's proximity to the person conducting the search, your farm's relevance to the searcher's intent, your ratings and reviews on Google and other sites, online mentions, and links to your site.

Get links to your site

Online directories (especially farm, PYO, or tourism sites that list farms) are one of the easiest places to get found online.

Online directories—especially farm, PYO, or tourism sites that list farms—are one of the easiest places to get found online. Most of these directories offer free listings. Some directories require memberships or payment to be included; if your target customers are likely to use a directory, paying for a listing can be worth it.

There are also general business directories (Better Business Bureau, Manta, YP.com, Yahoo), city or local directories (chambers of commerce, city websites), and *Buy Local* sites that can help your customers find you.

Also try to get listed on review sites, such as Yelp, TripAdvisor, and FarmStay, or on local Listservs or online forums. Search engines typically rank directories and reviews sites at or near the top of regular listings, so it's important to be included.

Look for ways to get links

One of the best ways to identify sources for web links and citations is to conduct an online search for your peers/competitors to see where they're listed. Some apps and browser extensions offer free tools that let you see exactly which sites are linking to your peers/competitors. Do an online search for "free competitor backlinks tool" to find one.

You can also contact sites to ask if they'd consider linking to your farm. Local businesses and organizations often promote vendors and other businesses they support. Most farmers markets list their vendors on the market websites and will link to your site.

Google particularly values links from *authoritative* websites, such as government agencies, Wikipedia, the *New York Times,* and other well-established and well-trafficked sites.

Direct links[46] to your site are especially helpful, but any mention of your farm (referred to as a *citation* in marketing jargon) is beneficial. Quality is important, but quantity is also helpful; if all other things are equal, having more links than your peers/competitors can give you an edge.

Post consistent and accurate information

When submitting a listing, make sure your farm's contact information (name, address, URL and phone) are spelled and abbreviated accurately and consistently. When you post accurate information, search engines and directories are more likely to find your farm and display your information correctly.

Free cnline tools can help you find all the places where your farm is listed online so you can easily correct inaccurate listings. Do an online search for "free citation audit" to find one.

Get found in Google's regular search

Appearing in Google's local listings is great, but Google only displays a Local 3-Pack when it's relevant. The rest of the time, your farm is vying with websites from any location to get a high ranking in search engine results. If there are many popular, well-established websites offering similar content, it will be difficult to rank at the top of Google's regular search results.

If there are many popular, well-established websites offering similar content, it will be difficult to rank at the top of Google's regular search results.

You may be able to improve your Google search rankings by using *search engine optimization (SEO)*. SEO encompasses a wide range of tactics; start with the basic principles discussed here.

Most of these suggestions will make your website more useful for your customers, so they're worth doing for their own sake. You'll also have a solid foundation if you decide to advertise, which we'll discuss later.

Warning: This next section gets a little wonky. If your eyes glaze over when you read technical stuff, skip ahead to **MEASURE.**

46. Marketers refer to links from external sites to yours as *backlinks* or *inbound links* to distinguish these types of links from internal inks from one page of your website to another page on your site.

Identify keywords and incorporate them into your website

The words that someone types into a search engine box are known as *keywords* or *keyword phrases*. In the web's early days, Google sought to match the exact word or words that a user typed into the search box with a web page using the same words. Today, Google searches for the underlying topic of a web page, not just keywords.

The words that someone types into a search engine box are known as *keywords* or *keyword phrases.*

This section provides basic tips on how to identify and use keywords. If you're planning to do extensive search engine optimization or online advertising, you'll need to conduct more extensive keyword research.

How to identify your keywords

Enter a keyword phrase (such as "alpaca farm") that your target customers are likely to use to search for your farm or offerings into the Google search box.[47]

Scan the regular listings (skip past the Local-3 Pack and map if shown). Does your farm appear in the first page of the search results or are you on page 5? Note *which* page of your website appears in your listing; it may not be your home page.

Continue viewing the search engine results for your keyword phrase. Google often displays two useful features in search results.

- The "People also ask" box shows a list of related search questions. Are there any keyword phrases in the list that would be a good match for your farm? If so, conduct the same search for that phrase to see where your page ranks.
- Google also usually displays a "related searches" box at the bottom of the search engine results page. Scan through this list to look for other keywords your customers might use. Google each appropriate keyword phrase to see where you rank for that query.

Review the possible keywords for each page of your site. Choose the phrase that gave you the highest search engine ranking and assign it to that page.

47. Google personalizes search results based on your location and previous searches, so you may not see the same results as someone searching for your farm for the first time. If possible, use a public computer for this research.

How to incorporate keywords into your web content

Use your keyword in your main headline. Placing your keyword in your headline signals to Google that your page includes relevant content for that keyword.

Tag your main headline as an "H1 headline" in HTML code. (Don't worry, you don't need to know how to code. Most website building software will provide a slot to enter your headline and automatically add the tags for you.)

If you can't find a natural way to include the keyword in your primary headline, include it in a subhead. (Better yet, use the keyword in both places if you can.) Subheads, or, secondary headlines, are known as H2 headlines in HTML code.

Use your keyword in your first sentence. Once again, look for a natural way to introduce your topic using your keyword. For example, you could say, "Our PYO apple orchard attracts thousands of visitors each summer and fall." If your keyword doesn't sound right in your first sentence, use it somewhere in the first paragraph.

Use your keyword (as well as close synonyms) multiple times. For example, your second paragraph could begin, "Apple picking (close synonym of PYO) has been a popular family activity in the Northeast for two centuries." Your third paragraph could begin, "Since [our orchard] began offering PYO apples in 1965, we've added a variety of other PYO fruits, including strawberries, blueberries, raspberries, peaches, and Asian pears."

Avoid *keyword stuffing*. While it's a good idea to use the keyword and related words a number of times on your page, it should flow naturally. If Google detects that you're artificially *stuffing* the keyword into the page for SEO purposes, it could penalize your site.

Create page titles and meta descriptions for your web pages

A *page title* (aka a *title tag*) is a string of words that name or suggest the purpose of a web page. A *meta description* briefly describes or promotes the content that a web page contains.

Page titles and meta descriptions get displayed when a web page is listed in Google search results. They usually don't appear on your website itself.[48]

---

48. Page titles *do* get displayed in browser tabs when someone is visiting that page on your website.

While page titles and meta descriptions only play a small role in determining your search engine rankings, they can play a major role in driving traffic to your website.

While page titles and meta descriptions only play a small role in determining your search engine rankings, they can play a major role in driving traffic to your website.

Let's look at how Google displays website listings on search engine results pages. "How the 'Six Ps' of Marketing Can Help Increase Sales" is one of the most popular pages on my website. This blog post has ranked at or near the top of Google results for certain keyword phrases for the past four years.

Google search results display at least three components for each listing:

- **The URL for the page** (in this case, https://goodeggmarketing.com/six-ps-marketing-increase-sales/).
- **The *page title*** (the title that appears below the URL).
- **The *meta description*** (a short bit of text that appears below the page title).

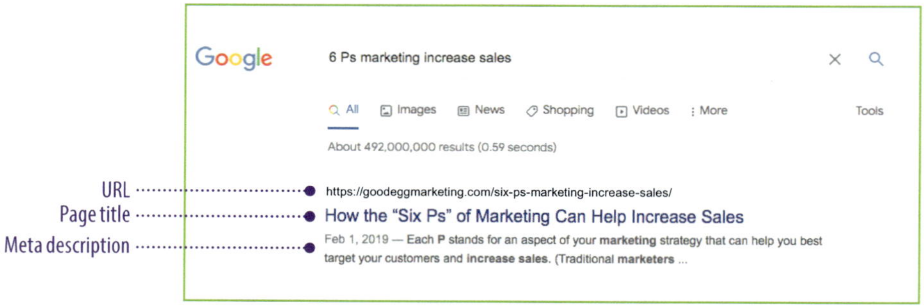

Screenshot of search engine results for the query "6ps marketing increase sales."

The words in the Google search box ("6ps marketing increase sales") can all be found in the page title.[49] Although the search box phrase used the numeral 6ps, while my page title spelled out the word "six," Google recognized that the two terms mean the same thing. Google also recognized two of the keyword phrases from the search box ("marketing" and "increase sales") in the meta description and bolded them to further highlight the match.

49. The page title you write to appear in search engines does not have to use the same words as the headline on your web page; in this case, I used the blog headline as my page title because I thought it was effective.

How to create an effective page title

Good Egg Marketing | Expert Marketing for Good Food & Good...

Make your page title easy to read in a glance.

Keep your text under 50–60 characters. If your page title is too long, the browser will cut of the text with an ellipsis (. . .) The exact number of characters that will fit before the ellipsis appears varies depending on the width of the characters in your typeface and the device it's displayed on.

You don't need to use complete sentences. If you're linking several unrelated pieces of information together, separate them with a hyphen (-), colon (:), or a pipe (|).

Edit out short words, such as "the," "is," "a," or "are." You can fit more words and it will be easier for readers to scan your page title.

Use your keyword phrase at or near the beginning of the page title. The words at the start of your page title will get the most attention and search engines assign more weight to the words at the start of the page title. For example, if your keyword phrase is "winter meat CSA," write, "Winter Meat CSA | Grass-fed beef & chicken," not "Join our grass-fed beef & chicken winter CSA." Putting your best stuff first is also helpful, in case your page title gets cut off in the browser.

Place your farm name at the start of the page title for your home page. If you have room, add enticing details. For example, the page title for my website starts with the website name/business name, followed by my tagline. Although the title's cut off, users can see the most important words

This close correlation between the search engine query and the words in the page title and meta description may explain why this blog post continues to rank so highly. Other factors, such as the fact that my site has been around for over a decade, has many sites linking to it, and has internal links, may also help. While not many people search for this phrase, those that do are very likely to see the listing and click on my post.

Page titles and meta descriptions must be embedded in the HTML source code for your pages for them to appear in search engine results. As I said before, most website platforms will let you enter the text in plain English and add the appropriate HTML tags for you.

How to create an effective meta description

Stick to under 120–155 characters. If your page title's too long, the browser will cut off the text with an ellipsis (. . .). The exact number of characters that will fit varies, depending on the width of the characters in your typeface and the device it's displayed on.

Incorporate your primary keyword phrase and related keywords for the page.

Use active language and compelling details. For example, your farm store page could state, "Order holiday pies made from our ripe fruit." Your CSA page could say, "Choose from over thirty vegetables, picked daily. Over 90% of our CSA members renew each year."

SEO also includes *usability* factors

Google wants the people who use its search engine to have a good experience, so they'll keep coming back. It wants to refer people to websites that provide a good *user experience.*

Here are key usability factors that impact SEO:

- Have compelling content
- Be easy to navigate
- Load quickly
- Be mobile-responsive
- Be secure

If two sites are equally relevant, but one is more user-friendly, Google is likely to give that site a higher search ranking.

See the *Grow your website* section for more details on usability factors.

Online Advertising

Online advertising can be the fastest way to achieve your goals. SEO can be a wonderful tool, but it takes time to learn and doesn't always work. Online ads get posted instantly and can have an immediate impact on your farm.

Most digital channels offer advertising and sponsorship opportunities. Your options include social media advertising (Facebook, Instagram, YouTube, LinkedIn, etc.), review sites (Yelp!, Trip Advisor, The Knot, etc.), Amazon ads (if you're selling online), Google Ads (search, display, shopping, and other ads), native ads on newsletters, and more.

Pricing for online advertising ranges from ultra-affordable to sky-high. You can advertise on Facebook for as little as $1 a day. How much bang you'll get for that dollar depends on a variety of factors.

Pricing for online advertising ranges from ultra-affordable to sky-high.

This section focuses on Facebook/Instagram and Google search ads, since those are the channels that farms are most likely to use. They also tend to be the easiest, most effective, and least expensive way to advertise online. Advertising via other channels can be effective, too, but each requires its own expertise.

There's plenty of free advice available to help you get started with online advertising. If you're serious about advertising, hire a consultant or agency to work with you. When working with an agency, expect to pay a minimum of $500 a month for the ads themselves-,and an additional $150–$500 a month for ad management.

General tips for online advertising

Be clear about your goals. Are you trying to sell holiday pies? Attract PYO customers? Cultivate CSA customers? Woo back former customers? Or are you just trying to get more of your fans to see your posts? Your choice of channel, type of ad, and content will depend on what you're trying to achieve.

Be clear about your target customers. What are they searching for online? Where are they in the buying process? What motivates them? How can your ad directly or subconsciously appeal to their values, needs and desires?

How to create a landing page

A landing page is a single webpage that visitors reach by clicking a link from an online ad, social media post, QR code, or other location. It can be a new or existing page on your website or a standalone page you create solely for marketing purposes.

Landing pages are used to promote a specific topic or ask, such as an event at your farm or CSA registration. A good landing page is simple and compelling.

Landing pages are often used for a short period of time and then deleted. They don't need to be added to the navigation menu, since the only way that people will see the page is if they click on the link you provided.

Set up your website to support your ad campaign. If you run an ad about an event, your CSA, or any specific offering from your farm, your ad should send them to the most relevant page for that offering. If you don't have a dedicated page on your website for it already, you should create a *landing page* that follows up on your ad content. Landing pages will make your ads more effective and can reduce your advertising costs.

Developing a successful ad campaign involves trial and error. Test, adjust, and try again.

Test, adjust, and try again. Developing a successful ad campaign involves trial and error. Test every variable, including your ad content (headline, images, text, and offer), your audience (who will see the ad), the logistics (budget, repetition, time of day, etc.), and other elements to achieve the best results.

Allow time for it to work. Don't run an ad once and expect immediate success. People usually need to see your ad multiple times before they respond.

Make sure it's paying off. The beauty of online advertising is that you can measure whether it's achieving your goals. Use the tools that each channel offers to track and measure the results.

Advertising on Facebook or Instagram

Meta (formerly known as Facebook) owns Facebook and Instagram, so it's easy to set up ads to run on both channels. But don't use exactly the same strategies and ads for Facebook and Instagram. Fans expect different things from each channel, so you're better off creating unique ads for each.

Both Facebook and Instagram offer two ways to advertise: *Boosting posts* and creating ads. A *boosted post* is content that you've already posted on your page that you can pay to get shown to current followers or new fans. An *ad* is usually created from scratch with a specific goal in mind.

Some marketers feel that boosting posts is a waste of money, but it can be an effective way to reach new followers and keep current followers engaged. Creating ads from scratch, however, gives you more choices and control over your content, audience, and where the ad is placed.

Some marketers feel that boosting posts is a waste of money, but it can be an effective way to reach new followers and keep current followers engaged.

Advertising costs vary depending on how you target your ads. Facebook works on a bidding system, so if the specific audience you're targeting is in high demand, it'll cost more to reach them. There's more competition at certain times of year (such as during the holidays), so your dollars won't go as far if you advertise then.

Google Ads

Google Ads (formerly known as Google AdWords) are text-based ads that appear in search engine results pages at the top or bottom of the organic listings. You bid to get your listing displayed when someone enters a specific keyword or keyword phrase into the search engine box.

When you pay for an ad. your listing can leapfrog to the top of Google's regular search results or show up above the Local 3-Pack. You can also choose to have your ad display on websites, apps, or videos (Google also owns YouTube).

How to boost a Facebook or Instagram post

Boosting posts is the easiest way to start advertising on Facebook, since you don't have to worry about creating new content. Facebook recommends that you boost a post that's already gotten engagement, since it's already been proven to have some appeal.

You can boost a post directly from your Facebook page by clicking the blue *Boost post* button. Facebook will ask you to choose a *Goal,* a *Call to Action* button (such as *Learn More, Shop Now, Book Now, Sign-up,* or *Send Message*), and an *Audience* (the types of people you want to see the ad).

The ability to target specific audiences is one of Facebook's best features. Facebook has a wealth of information on each user's demographics, interests, and behavior, so it can show your post to the people most likely to be interested in your farm. You can let Facebook identify a *Smart audience* using artificial intelligence or select *People you choose through targeting.*

I recommend that you select your own audience, so you can specify the location, age, gender and interests of people you want to reach. Browse the options to find topics that are likely to interest your target customers. While you can't specifically target *locavores,* you can show your post to people with an interest in organic food or gardening.

Once you've set your audience, Facebook will ask you how long you want your boosted post to run and how much you want to spend. Facebook will tell you how many people you'll be able to reach by changing your budget or time period or expanding (or narrowing) your audience.

You can choose to run an ad for a few days and extend the schedule if you sense that it's working.

Your final step is to choose where you want the boosted post to appear. Boosted posts can appear in your Facebook News Feed, Instagram, and/or Messenger. Stick with the channel where you created the ad (e.g., if it was a Facebook post, boost it only in Facebook), at least to start.

You only pay when a user takes a specific action, such as clicking on your ad to visit your website or call your farm.[50] The average *cost-per-click* typically ranges from $1 to $2, but it varies by industry and term. For example, the average cost for a single click in the insurance industry was $20.12 in 2021, compared to $0.77 per click in the electronics industry.[51]

Another advantage to Google Ads is that it's fairly easy to track and measure your results if you have Google Analytics installed on your site. For example, if you're running Google ads for your CSA, you'll be able to track how many of the people who clicked on the ad ultimately purchased a share. (If you're using Google Display ads, you can set up a *Cost Per Acquisition* campaign where you only pay when someone makes a purchase.)

It's simple to set up a Google Ad, but hard to manage all the variables. Google offers an automated *Smart campaigns* option to set up your ads. If you don't understand how the system works, it's easy to waste your money. You have to identify the most profitable keywords for your farm, develop content that attracts your target customers, know how to interpret results, and be able to recalibrate when necessary.

It's simple to set up a Google Ad, but hard to manage all the variables.

Google estimates that businesses generally make about $2 in revenue for every $1 they spend on Google Ads.[52] It may take a substantial budget to run an effective test. Unlike Facebook Ads, where you can sometimes see results from running a very small test ad with a budget as low as $50, it may take hundreds or even thousands of dollars to determine whether Google Ads work for your farm. If you

50. You must set up a dedicated phone number to track calls from the ad to use this feature.

51. Statistica. (2022, July 11). *Average cost-per-click (CPC) in Google Ads search advertising in selected industries in the United States in May 2021 (in U.S. dollars).* https://www.statista.com/statistics/1115432/us-search-advertising-cpc/

52. Google Economic Impact (2022). *How we calculate Google's economic impact.* https://economicimpact.google.com/methodology/

do it on the cheap, you might bid too low or choose lower-cost keywords that aren't as effective. While it's hard to generalize, most experts say you need to invest a minimum of $150–$500 a month to run an effective test.

It usually takes time to get it right. There's a lot of trial and error involved in choosing the right keywords, crafting the right text, choosing the right goals, making the right bids, etc. You have to allow time to get initial results, interpret them, adjust your campaign, and evaluate the results. And Google needs a certain volume of clicks over some time for its machine learning to kick in. While you may be lucky enough to get an immediate response, experts say to allow six weeks to three months to see if it works.

I'm skeptical about whether Google Ads is a good idea for most farms. Unless you're in an extremely competitive market, you're better off investing your time and money keeping an active Google Business Profile, adding relevant content to your website, or using other types of advertising.

If you're serious about Google Ads, I recommend that you work with an agency or consultant, but be prepared to pay. Most experts will charge you at least $500 a month for managing your Google Ads (not including the cost of ad placement).

MEASURE

"Half the money I spend on advertising is wasted; the trouble is I don't know which half." JOHN WANAMAKER

TOPICS

- Is your marketing working?
- Measurement principles for all size farms
- Measuring sales and display tactics
- Measuring your Google Business Profile
- Measuring print materials
- Measuring traditional advertising
- Measuring email
- Measuring the effectiveness of your website
- Measuring free and paid online marketing

What you'll find in MEASURE

Where to focus

MICROFARM MUSTS
Use simple measurement techniques to learn whether your road signs, Facebook page, road signs, farmers market display, or other tactics are helping people find your farm.

SMALL FARM SMARTS
Use analytics to evaluate whether your website, social media, emails, and other tactics are helping you achieve your goals.

MIDSIZE FARM MAYBES
Use your marketing strategy to plan which tactics to use. Test each tactic to find out which works best for your farm.

This part of the book introduces you to principles and techniques for measuring your marketing tactics.

Is your marketing working?

Marketing requires time, effort, money, and other resources. How can we tell if we're using the right tactics to achieve our goals?

Measurement principles for all size farms

Here are a few things to keep in mind when you're deciding how to measure any marketing tactics.

Measurement techniques

This section guides you through how to measure all the marketing tactics covered in this book, including sales and display tactics, your Google Business Profile, print materials, traditional advertising, email marketing, your website, and free and paid online marketing.

Is your marketing working?

John Wanamaker (1838–1922) was the founder of one of the first department stores in the US. Known as a savvy marketer, he was a big believer in advertising. Wanamaker famously said, "Half the money I spend on advertising is wasted; the trouble is I don't know which half."

Even though the department store magnate had plenty of money, he didn't want to waste it on anything that wasn't helping to achieve his goal (making more money).

Measurement is a means to an end, not a goal in itself. The purpose of tracking, measuring, and evaluating your marketing tactics is to figure out whether the time and/or money spent on marketing is paying off—and if not, what to do about it.

Marketing encompasses all the 6 Ps: Product, Price, Placement, Presentation, People, and Promotion. Since most farmers are thinking about promotion when they wonder if their marketing's working, this section primarily focuses on how to measure promotional tactics. These tactics include conventional activities (distributing flyers or holding events), as well as online tactics (social media or email marketing).

Today's marketers have more measurement tools at our disposal than Mr. Wanamaker had, yet figuring out what's working is still challenging. Some tactics are just harder to measure than others. For example, it's difficult to pinpoint whether traditional tactics like handing out flyers or running radio ads are working, yet you can easily track an online sale to a specific email. Even simple forms of tracking, measurement, and evaluation can help you make better decisions about how to use your marketing time and money, so it's worth investing a little time learning how to use them.

Measurement principles for all size farms

Decide what you'll test and how you'll measure it

Before you choose a tactic, ask yourself:

- Which tactic do I want to test or evaluate?
- If it succeeds, how will it benefit my farm? What business or marketing goal do I want to achieve by using this tactic?
- Which techniques or tools will I use to measure the results or impact of the tactic?

After you launch the tactic, ask yourself:

- Does the value of the results exceed the time, money, and effort it took to implement the tactic? If so, do I want to continue?
- (Advanced) How does the value of this tactic compare to other marketing tactics?
- (Advanced) Do I want to test any variables (text, images, offer, placement) to see if they improve the results?

Keep records

Record your baseline. Before you make a change in any aspect of your marketing mix, measure and write down the metrics you want to track. For example, if you want to measure whether a fresh merchandising display increases sales of a particular product, record the sales for that product before you set up the new display, so you'll be able to compare sales before and after the change.

Log any external factors that could impact your results. Keep a digital file or notebook where you can jot down anything that could affect sales or operations, such as weather conditions, media coverage, or the opening of a new farm down the road. These external variables may have a larger impact on your results than the specific tactic you're evaluating; if so, you may need to test your tactic again. Your records will be especially useful when comparing year-over-year sales. If this year's July sales jumped or dropped, you'll be able to check the history for a potential explanation.

Use existing data. If you can access customer data from your POS system or loyalty program, you'll have a treasure trove of information to help evaluate your tactic's success. Even without these tools, you and your staff are, consciously or subconsciously, collecting information and making observations when you interact with customers, so make a habit of recording these insights.

Use analytics tools. Social media channels and email service providers offer built-in analytics that can be useful. Your website platform may provide analytics, or you can install tools such as Google Analytics and Google Search Console. If you do any paid advertising, you'll have access to extensive analytics to help you evaluate results.

Do the minimal amount of measurement necessary

- If an easy, inexpensive new marketing tactic seems like a no-brainer, go with your instincts, even if you don't have any data to back it up.
- Don't waste time measuring results unless you can use the information you collected to make better business decisions.
- Put most of your effort into measuring your newest, riskiest, most time-consuming and/or expensive tactics.

Don't waste time measuring results unless you can use the information you collected to make better business decisions.

Collect enough data to make reliable conclusions

If your farm or the offering you want to measure is fairly new or you serve a small number of customers, you may not have enough data for a definitive analysis. For example, if you start a CSA and have 25 members your first year and 20 of them renew for a second year, your retention rate for year 2 is 80%. While that retention rate is currently accurate, you don't have enough of a track record to know if it will remain at 80% year after year.

Similarly, if your 200 Facebook fans are mostly friends and family who comment on everything because they want to help you, you can't assume that if you acquire 200 new Facebook fans who are strangers, you'll get twice as many comments, shares, and Likes.

Some hard-to-measure tactics are still worth using

It's easier to measure quantitative outcomes—volume sold, average sales per customer, or your CSA retention rate—than qualitative goals, such as "Increase brand awareness." Just because something is hard to measure doesn't mean it isn't worth doing.

If sales jump after you post a new video, it's reasonable to assume that the videos played a role, even if you can't prove it.

If you post some funny videos about your farm, you can measure how many people view the videos. You won't know if someone became a customer after watching the video unless you ask them or they tell you. If sales jump after you post a new video, it's reasonable to assume that the videos played a role, even if you can't prove it.

Measuring sales and display tactics

Increasing sales at the farmers market, your farmstand, or other settings may require some new marketing tactics, such as enticing people to stop at your booth or changing how you bundle and price your items.

Here are some techniques you can use to evaluate whether your tactics are making an impact.

Observation

- Are more people stopping at your booth? Are they looking at whatever you've changed? Are they staying longer?
- Pay attention to facial expressions, body language, and tone of voice. Do customers seem excited, interested, happy, or are they impatient, agitated, upset?

Conversation

- Talk to your customers. If you're trying to measure the impact of a single change, ask everyone the same question: "Have you stopped at our booth before?" Keep a clipboard or digital file open where you can tally the responses.
- Listen to your customers. What questions are they asking? What are they looking for?
- Eavesdrop. What are they saying to their companions?

Counting

- Use a hand tally counter clicker to count how many people stop by.

Social media

- Look for customer posts that mention, comment, or react to the change. Post a photo of the change and ask for comments.
- Conduct a poll on Facebook, Instagram Stories, Messenger, or another channel.

Email

- Pose a question or run a poll/survey to get feedback.

Analyze online reviews

- Read online reviews for your farm and look for mentions of the change.
- Run a free *word cloud generator* app to see which words get mentioned the most.

Measuring your Google Business Profile

Google provides data on how people are interacting with your Google Business Profile (GBP). There are a variety of ways to access this information. The easiest way to view it is to read the performance report that Google emails you each month. You can also open your profile (click Promote and then Performance).[53]

Your report shows how many people viewed your profile and subsequently clicked on your profile to call, message, get directions, or visit your website. Check to see if monthly interactions are increasing over time.

You can also see how many times your profile was displayed when a user searched for a specific keyword or phrase and a list of the keywords/phrases used. Scroll through the list to look for keyword ideas. The most useful phrases are likely to be near the bottom; click "see more" to view the full list.

---

53. In my opinion, Google makes it ridiculously difficult for you to view and download this information. You can download some of the data if you access your profile through Business Profile Manager. Visit https://support.google.com/business/answer/9918094 for instructions.

Google only displays data for the past six months. If you want to track your performance over a longer period of time, set up a spreadsheet so you can manually enter and save the data.

You can try to increase the number of interactions with GBP by keeping your profile up-to-date, adding photos, and encouraging/responding to reviews. If you're serious about increasing your ranking in Google's local or regular search, you'll need to invest in search engine optimization for your website.

Measuring print materials

It's difficult to measure the impact of print materials unless you set up a mechanism to track how they're used.

People tend to discount the value of print materials in our digital age; print isn't dead, it's evolving. You don't need lengthy brochures or newsletters; print postcards, rack cards, or flyers that share a quick message and direct people to your website or social media. Color printing is much more effective at getting attention than black-and-white; keep costs down by printing color copies at home or finding good deals at printshops. Save money—and trees—by printing fewer copies; create a display copy of your printed piece and encourage people to snap a photo or scan a QR code to capture the information.

It's difficult to measure the impact of print materials unless you set up a mechanism to track how they're used.

- If you place a stack of flyers at a local coffee shop, you can return the next week to see if the pile is smaller, but you won't know if the flyer prompted anyone to visit your farm.
- If the flyers you place at the coffee shop promote a specific item ("Order holiday pies") and you get more orders than usual, the flyer probably generated some orders.
- If the flyer you place at the coffee shop include a discount coupon for holiday pie orders, you can track how many people redeem the offer.
- If you print the holiday pie discount flyer in different colors, you can distribute a unique color to each location, so you can track which setting generates the best response.

QR codes have become a good tracking mechanism. These codes were initially developed in 2004, but didn't gain consumer acceptance until the COVID-19 pandemic. Create unique QR codes for each setting, so you can track responses by channel, local, and/or offer.[54] If you don't have a website, you can set up a QR code to send people to your social media page.

You can track direct mail (printed materials mailed to residential households) by including a unique coupon or CTA for each campaign, channel, or location. You can also ask your printer to use *variable data printing* to individualize each piece with a customer's name or unique promo code, so you can track responses by household.

Offering discounts, free gifts, or using direct mail can be expensive. Calculate the additional sales generated by the promotion and subtract your expenses to see if it generated a profit. If possible, also measure the long-term impact of the promotion; if some of the people you attracted become long-term customers, the investment may pay for itself over time.

Your marketing strategy should include some tactics that increase awareness of your farm or build good will, even though you can't measure whether they have a direct impact on one of your goals. For example, people love *swag*—a refrigerator magnet, a chicken stress ball, or a packet of seeds. If you give your best customers a fun and useful item imprinted with your farm name, they'll be subconsciously reminded of your farm when they see or use it.

Your marketing strategy should include some tactics that increase awareness of your farm or build good will, even though you can't measure whether they have a direct impact on one of your goals.

Measuring traditional advertising

Research shows that traditional advertising (print, radio, and TV) tends to be more effective at creating brand awareness and attracting new customers than social media.[55] Although traditional advertising is usually more costly than digital advertising, some local outlets can be very affordable.

54. You must have Google Analytics or another web analytics program installed in order to track which QR code generated a click to your website.

55. Bettencourt, L. (2018). "Is Shifting from Traditional Advertising Good for Business?" *American Marketing Association*. https://www.ama.org/2018/03/21/is-shifting-from-traditional-advertising-good-for-business/

Like other brand awareness efforts, it can be difficult to track the impact of your ads. The advertising outlet can tell you how many people were exposed to the ad, yet there's no guarantee that the viewers were in the room when the ad aired.[56] If you have a surge in sales, visits, phone calls, and/or web traffic right after your ad runs, you can credit the ads.

It's easier to measure the impact of advertising if you're promoting an event or specific offer. When possible, test one offer or outlet at a time. Include a coupon or promo code, provide a unique phone number to call, or send viewers to a unique URL to facilitate tracking.

Measuring email

Email is a valuable way to communicate with your customers and help them feel connected with your farm. It's also a wonderful tool for promoting products, events, and other offers.

Your email list is a business asset. Email is a valuable way to communicate with your customers and help them feel connected with your farm. It's also a wonderful tool for promoting products, events, and other offers. While sending emails is free or inexpensive, it takes a lot of time to create high quality messages, so make sure that you're getting as much value as you can from your email program.

Study the response to each message you send. Which subject lines get the most people to read the newsletter? Which newsletter topics are the most popular? Are the number of subscribers growing? How many people are clicking on links in your emails to learn more, RSVP, or make a purchase? Use your observations to test and refine your email program.

When it comes to measurement, most of my clients rely on the data that their email service providers (ESPs) like Constant Contact and Mailchimp provide—especially the open rate. ESPs calculate your open rate by taking the number of people who opened the email and dividing it by the number of emails that were successfully

56. Most forms of advertising are sold based on number of impressions, meaning the number of times that an ad was broadcast or aired to a person. Impressions are sold by cost per thousand (CPM). In theory, this makes it easy to compare the relative cost of advertising online, running a billboard, doing TV ads, etc.

delivered. If your emails have an average open rate of about 50%, that means roughly half of your list are reading your emails.

Open rates have never been reliable and are now likely to be inflated.[57] While you should continue to watch your open rates, focus on how many people click links in your messages. If you post a link to register for an upcoming event, track how many tickets you sell in the next 24 hours. If you run a story about one of your offerings, track your sales before and after to see if sales increase.

Email can be used as part of an overall strategy to attract new customers. Use social media or advertising to entice people to join your email list. Once they subscribe, send them a series of welcome emails, encouraging them to join your flower CSA, come for PYO, or whatever goal you're trying to accomplish.

Measuring the effectiveness of your website

If you've invested resources into creating a website, take a little time to figure out whether its helping you to attract customers, sell your offerings, educate visitors, and other goals.

Many of your online marketing tactics are tied into your website. For example, your email subscription box is probably embedded in your website. When you evaluate the effectiveness of your website, you'll also be able to look at how well your other marketing tactics are working.

Most website platforms provide built-in analytics that answer some of these questions. If you want more in-depth information, install a third-party web analytics program.

---

57. ESPs insert an invisible tracking pixel into each email; if the pixel pings the ESP, it gets counted as an open. Spam, image blocking, and other factors can block or over-count the pixel, distorting the true open rate. Apple Mail disabled pixel tracking for privacy-protected users in September 2021; their emails get reported as opened, regardless of whether they've been viewed. Since approximately half of all email goes through Apple Mail, your open rates are probably a lot lower than reported.

How to choose a web analytics program

Web analytics tools measure how well your website is helping you reach your goals. Most basic analytics programs will report the number of monthly visitors, how visitors got to your site, how long they stayed, and the most popular content.

More sophisticated analytics programs enable you to trace how many web visitors from each source (email, Facebook, digital ad, link from another site) took a specific online action (join your CSA, purchase from your online store). This helps you identify the best tactics.

For the past decade, I've enthusiastically recommended Google Analytics (GA), a free, powerful, analytics program. The newest version, GA4, has many improvements, but it's too hard for non-technical people to learn.

Most website platform offer basic analytics for your website; that may be all you need. If you want a more robust program that's still easy to use, try one of the free or inexpensive alternatives to GA. As of this writing, I've heard strong recommendations for Matamo and Piwik. Do a web search for "GA4 alternatives" to learn more.

If your farm advertises, sells, and markets online, GA4 is still the best web analytics program out there—and it's free. Hire a consultant or agency to set it up and train you to use it.

Website effectiveness metrics to review

- What are the most popular pages on your site?
- What percentage of your traffic comes from each social media channel or from email?
- What are the top sites that are referring people to your site?
- What search terms do people use to get to your site?
- How many unique individuals visited your site last month?
- How much time does the average user spend on your site?

If you're selling online or running online ads, install a robust analytics program like GA4 that enables you to test and evaluate every aspect of your program.

Suggested Ways to Measure Free Online Tactics

GOAL	TACTIC(S)	MEASUREMENT
Increase sales at your farmers market booth, farmstand, or other on-farm, in-person activities	Promote a specific item, service, event, or activity on one social media channel, email, or other online outlet	Measure whether sales of the promoted offering increase
Attract new customers and increase brand awareness	Add categories, information, and posts to your Google Business Profile that emphasize the keywords people are likely to use when doing a Google local search Do an online search for other farms to see which online directories and websites link to them; try to get your farm listed in the same places	Record your farm's ranking in Google's local search results; track it to see if it improves over time Use Google Business Profile analytics to measure phone calls, clicks for direction, and other engagement to see if volume is increasing
Deepen customer engagement	Cultivate a small but engaged base of online supporters who will comment on and share your posts, write online reviews of your farm, and click email links to read articles on your website[58]	Use built-in analytics from each channel to see if your engagement rates increase Compare engagement rates on different types of posts and focus on the ones that get the best response Measure whether the number and quality of your online reviews increase Measure whether the number of clicks on email links increase

Measuring free and paid online marketing

Measuring free online marketing tactics

"Free" online marketing tactics aren't truly free, because it takes time for whoever is doing the work.

What's the value of acquiring a Facebook fan, attracting a new email subscriber, or getting an online review? It depends on your farm's size, goals, and business model, as well as how skillfully you can use these tactics.

---

58. Engagement rates are based on the percentage of your fans who interact with your posts, not the size of your fan base.

If you use a variety of unpaid marketing tactics to promote your farm, it can be difficult to determine which specific tactics are worth your time. Try increasing your efforts on one tactic for a few months to see if it has any impact. For example, you could run a weekly recipe contest on Facebook to see if you get more engagement. If you don't see any change, focus more of your attention on a different tactic.

See the *Suggested ways to measure free online tactics* chart on the previous page for measurement ideas.

Measuring paid online marketing tactics

If you're serious about attracting new customers, increasing sales, and deepening customer loyalty, invest in some paid online marketing tactics, such as boosting Facebook posts, running Google search ads, or advertising on relevant podcasts and enewsletters. To maximize your investment, clarify your goals, their value to your farm, and how you'll measure the return on your time and money before you begin.

There are more options for paid online marketing than ever! Ask your current customers which social media, radio stations, online directories, and podcasts, they pay the most attention to.

Most advertisements must run repeatedly and over time for your target audience to recognize and respond to your messages. Maximizing the effectiveness of your ads requires expertise, skill, and testing/trial and error.

Every aspect of your ad can impact the results—including placement, budget, targeted audience, keywords, format, creative content (design, headlines, copy, and images), and offer. Some outlets offer automated advertising that will determine which combinations work best, but you typically need a large advertising budget for such tests to work. Whether you test your advertising manually or use the automated options, it can still take weeks or months to optimize the results.

Suggested Ways to Measure Paid Online Tactics

GOAL	TACTIC(S)	MEASUREMENT
Increase sales at your farmers market booth, farmstand, or other on-farm, in-person activities	Run a boosted post or ad to promote a specific item, service, event, or activity on a single social media channel, email, or other online outlet Offer a discount or promotion only available with a mention of the ad	Measure whether sales of the promoting offering increase Track how many customers claim the discount Compare which channels and ads generate the best responses
Attract new customers and increase brand awareness	Boost posts or run targeted ads to attract likely customers Promote a single offering or unique discount in an ad Add content to your website on niche topics likely to be of interest to your target customers Sponsor local events or partner with local businesses and organizations to include links to your website	Track overall sales, sales of promoted offerings, and number of customers to see if numbers increase. Compare return on investment for each outlet or type of promotion If you sense someone is a new customer, ask how they heard about you Check to see if your farm's ranking increases in Google's regular search results
Deepen customer engagement	Invest in developing video, Stories, and other tactics that entice engagement Boost your most popular Facebook or Instagram posts so that they get shown to more of your existing fans. Message everyone on your email list asking them to click if they want to continue to subscribe. Move people who don't respond after two tries to a list of inactive subscribers.	Use built-in analytics from each channel to analyze engagement rate Compare engagement rates on different types of posts and focus on the ones that get the best response Compare the number of email opens and clicks you get before and after purging inactives

How can you tell if your paid online tactics are working?

If you sell anything (an ecommerce product, a CSA share, or a ticket for an experience) directly from your website, social media, or any online channel, you'll be able to track exactly where each sale comes from. If the revenue that you generated exceeded the cost of posting or advertising, your tactic worked.

If you're using paid online tactics to persuade someone to take an action in real life (such as buying your cheese plate of the week at a farmers market), it can be difficult to prove which tactic resulted in a sale (unless you include a coupon, discount code, or a unique offer that can be tracked). Try to use one new tactic at a time, so you can look for indirect indicators that the tactic is having an impact. See the *Suggested ways to measure paid online tactics* chart for measurement ideas.

When evaluating the effectiveness of your paid tactics, try to determine whether that tactic generated a one-time sale or repeat customer. Attracting a repeat customer has much greater value for your farm. For example, a new customer might generate $100 in revenue in their first year and $200 the next year. Over two years, their value to your farm—referred to as their *customer lifetime value* (CLV) in marketing jargon—is at least $300.

If you have a POS or loyalty card system that tracks how many new customers you attract per year, the average revenue per customer, and the average number of years your customers have been with your farm, you'll be able to estimate the CLV of a new customer. If the cost of acquiring a new customer is less than the cost of paying for ads or other tactics, your tactics are paying for themselves.

Online advertising and other paid tactics can be profitable, but they don't work for every farm. If you decide to try paid tactics, budget a small amount and see what happens. Most online channels offer metrics that help you measure the performance of a paid activity.

Calculating whether paid tactics are profitable for your farm can be challenging. If you're serious about investing in online advertising, take the time to educate yourself or a staffer how to do it profitably or hire a professional to work with your farm.

HARVEST

"What we plant in the soil of contemplation, we shall reap in the harvest of action."

MEISTER ECKHART

Farewell, Farmer!

I hope you've enjoyed reading this book as much as I've enjoyed writing it!

Marketing is cumulative. Much of the value will come to fruition over time The more marketing you do, the more you can apply to next year's harvest.

Keep these principles in mind as you plan your marketing tactics.

Farm marketing isn't one size fits all. Just because the farmer down the road is a TikTok maven doesn't mean it will work for you. Choose tactics that make sense for your scale, goals and resources.

Marketing isn't just about your logo, social media, and advertising. When you're looking for the right tactics to achieve your goals, consider all 6 marketing elements: Products, Price, Placement, Presentation, People, and Promotion.

Prepare for both the known and unknown unknowns. Keep your eyes open for new product and technology trends.

Look for ways to partner and collaborate with other farmers and businesses who serve people with similar values.

Your purpose, values and brand personality are what differentiate your farm. Express them through your words, images, and actions.

I hope you've already planted at least one marketing tactic from this book and reaped the benefits.

Let's share some seeds. Please email me at hello@marketingyourfarm.com. Tell me what tactics are working for you. I promise to pass them along to other farmers.

I want you to thrive because your farm is essential to the world I want to live in. Please keep farming!

Photo: Deborah J. Karson

About the author

Myrna Greenfield is the *Top Egg* at Good Egg Marketing, a Massachusetts-based business that offers sales and marketing strategies, branding, and digital services for farms and food companies, small businesses, and non-profits. Greenfield hatched Good Egg in 2010 to promote good food and good causes. She's taught workshops and provided services to more than one thousand farmers.

Greenfield has worked in the food and beverage, marketing, health care, and high-tech industries. She was employee #1 at Equal Exchange, a worker co-op that sells fairly traded coffee, tea, chocolate, and other products from small farmers. The former director of Communications and Campaigns at Oxfam America, she has served at many other non-profits, including the Unitarian Universalist Service Committee, Fenway Community Health, and the Child Care Resource Center.

She's a frequent speaker and trainer at conferences, events, and meetings on topics such as agritourism, branding, sales and marketing strategies, websites, and social media. She's presented for a variety of groups, including the Glynwood Center for Regional Food and Farming, Harvest New England Agricultural Marketing Conference, National Farm to School Network, National Farm Viability Conference, New Entry Sustainable Farming Project, Northeast Organic Farming Association, Pasa Sustainable Agriculture Conference, and the South Dakota Specialty Producers Association.

Greenfield is a board member of Boston Farms Community Land Trust; Boston Farms turns vacant urban land into collective opportunity, supporting neighborhood farmers to grow the food that sustains us. She was a long-time board member of the Sustainable Business Network of Massachusetts and cofounded the Egleston Farmers Market.

She was raised in the Chicago area and has been a proud resident of Jamaica Plain, MA, for more than thirty years. She holds an MBA from Simmons School of Management.

Made in the USA
Las Vegas, NV
03 November 2026

57422082R00102